Springer Tracts in Advanced Robotics

Volume 140

Frank Chongwoo Park, Mechanical Engineering Department, Seoul National University, Seoul, Korea (Republic of)

S. E. Salcudean, The University of British Columbia, Vancouver, BC, Canada

Roland Siegwart, LEE J205, ETH Zürich, Institute of Robotics & Autonomous Systems Lab, Zürich, Switzerland

Gaurav S. Sukhatme, Department of Computer Science, University of Southern California, Los Angeles, CA, USA

The Springer Tracts in Advanced Robotics (STAR) publish new developments and advances in the fields of robotics research, rapidly and informally but with a high quality. The intent is to cover all the technical contents, applications, and multidisciplinary aspects of robotics, embedded in the fields of Mechanical Engineering, Computer Science, Electrical Engineering, Mechatronics, Control, and Life Sciences, as well as the methodologies behind them. Within the scope of the series are monographs, lecture notes, selected contributions from specialized conferences and workshops, as well as selected PhD theses.

Special offer: For all clients with a print standing order we offer free access to the electronic volumes of the Series published in the current year.

Indexed by DBLP, Compendex, EI-Compendex, SCOPUS, Zentralblatt Math, Ulrich's, MathSciNet, Current Mathematical Publications, Mathematical Reviews, MetaPress, and Springerlink.

More information about this series at http://www.springer.com/series/5208

Marco Tognon · Antonio Franchi

Theory and Applications for Control of Aerial Robots in Physical Interaction Through Tethers

Springer

Marco Tognon
LAAS-CNRS
Toulouse, France

Antonio Franchi
University of Twente
Enschede, The Netherlands

ISSN 1610-7438 ISSN 1610-742X (electronic)
Springer Tracts in Advanced Robotics
ISBN 978-3-030-48658-7 ISBN 978-3-030-48659-4 (eBook)
https://doi.org/10.1007/978-3-030-48659-4

This Springer imprint is published by the registered company Springer Nature Switzerland AG
The registered company address is: Gewerbestrasse 11, 6330 Cham, Switzerland

Series Editor's Foreword

At the dawn of the century's third decade, robotics is reaching an elevated level of maturity and continues to benefit from the advances and innovations in its enabling technologies. These all are contributing to an unprecedented effort to bringing robots to human environment in hospitals and homes, factories and schools; in the field for robots fighting fires, making goods and products, picking fruits and watering the farmland, saving time and lives. Robots today hold the promise for making a considerable impact in a wide range of real-world applications from industrial manufacturing to healthcare, transportation, and exploration of the deep space and sea. Tomorrow, robots will become pervasive and touch upon many aspects of modern life.

The *Springer Tracts in Advanced Robotics (STAR)* is devoted to bringing to the research community the latest advances in the robotics field on the basis of their significance and quality. Through a wide and timely dissemination of critical research developments in robotics, our objective with this series is to promote more exchanges and collaborations among the researchers in the community and contribute to further advancements in this rapidly growing field.

The monograph by Marco Tognon and Antonio Franchi is the outcome of the first author's Ph.D. thesis on tethered aerial vehicles. The work comes at the intersection of two hot research areas, namely aerial robotics and robot physical interaction, and constitutes a valuable reference in the new field of aerial physical interaction. A number of controllers and observers are proposed to fully exploit aerial robots' capabilities to interact with the environment, and even a multi-tethered system composed of two aerial robots is considered.

The techniques are effectively implemented in a large number of simulated and experimental tests, including a significant use case, and the various results are keenly discussed. A very fine addition to the STAR series!

Naples, Italy
April 2020

Bruno Siciliano
STAR Editor

To beloved family and friends.

Preface

This book focuses on the study of autonomous aerial robots interacting with the surrounding environment, and in particular on the design of new *control* and *motion planning* methods for such systems. Nowadays, autonomous aerial vehicles are extensively employed in many fields of application but mostly as autonomously moving sensors used only to sense the environment. On the other hand, in the recent field of *aerial physical interaction*, the goal is to go beyond sensing-only applications and to fully exploit aerial robots capabilities in order to interact with the environment, exchanging forces for pushing/pulling/sliding, and manipulating objects. However, due to the different nature of the problems, new control methods are needed. These methods have to preserve the system stability during the interaction and to be robust against external disturbances, finally enabling the robot to perform a given task. Moreover, researchers and engineers need to face other challenges generated by the high complexity of aerial manipulators, e.g., a large number of degrees of freedom, strong nonlinearities, and actuation limits. Furthermore, trajectories of the aerial robots have to be carefully computed using motion planning techniques. To perform the sough task in a safe way, the planned trajectory must avoid obstacles and has to be suitable for the dynamics of the system and its actuation limits.

With the aim of achieving the previously mentioned general goals, this book considers the analysis of a particular class of aerial robots interacting with the environment: *tethered aerial vehicles*. The study of particular systems, still encapsulating all the challenges of the general problem, helps on acquiring the knowledge and the expertise for a subsequent development of more general methods applicable to aerial physical interaction. This work focuses on the thorough formal analysis of tethered aerial vehicles ranging from control and state estimation to motion planning. In particular, the differential flatness property of the system is investigated, finding two possible sets of flat outputs that reveal new capabilities of such a system. One contains the position of the vehicle and the link internal force (equivalently the interaction force with the environment), while the second contains the position and a variable linked to the attitude of the vehicle. This shows new control and physical interaction capabilities different from standard aerial robots in contact-free flight. In particular, the first set of flat outputs allows realizing one of the first "free-floating" versions of the classical hybrid force-motion control for standard grounded manipulators.

Based on these results we designed two types of controllers. The first is an easy-to-implement controller based on a hierarchical approach. Although it shows good performance in quasi-static conditions, actually the tracking error increases when tracking a dynamic trajectory. Thus, a second controller more suited for tracking problems has been designed based on the dynamic feedback linearization technique. Two observers, for the 3D and 2D environments, respectively, have been designed in order to close the control loop using a minimal sensorial setup. We showed that the tether makes possible to retrieve an estimation of the full state from only an IMU plus three encoders for the 3D case, while from just an IMU for the 2D case. Parts of those results were extended to a novel and original multi-robots case as well. We considered a multi-tethered system composed of two aerial robots linked to the ground and to each other by two links. The theoretical results on generic tethered aerial vehicles were finally employed to solve the practical and challenging problem of *landing and takeoff on/from a sloped surface*, enhancing the robustness and reliability of the maneuvers with respect to the contact-free flight solution.

This work has been supported by the European Union's Horizon 2020 research and innovation programme under grant agreement No 644271 AEROARMS.

We would like to express our deep gratitude to Dr. Anthony Mallet for the excellent and continuous maintenance of the software and hardware framework. Without his precious work, the several experiments conducted during this work would not have been possible.

A sincere gratitude go to all the colleagues that contributed to this work: Sanket Dash, Andrea Testa, and Enrica Rossi.

Toulouse, France
February 2020

Marco Tognon
Antonio Franchi

Contents

Acronyms

ACADO	Automatic Control and Dynamic Optimization
AM	Aerial Manipulation
APhI	Aerial Physical Interaction
AR	Aerial Robot
AV	Aerial Vehicle
BLDC	Brushless DC
CoG	Center of Gravity
CoM	Center of Mass
DFL	Dynamic Feedback Linearization
DoF	Degree of Freedom
EE	End-Effector
ESC	Electronic Speed Controller
GPS	Global Positioning System
HC	Hierarchical Control
HGO	High Gain Observer
IMU	Inertial Measurement Unit
MoCap	Motion Capture System
SFL	Static Feedback Linearization
SLS	Standard Linear Solid
UAV	Unmanned Aerial Vehicle
VTOL	Vertical Takeoff and Landing Vehicle

List of Figures

List of Tables

Chapter 1
Introduction

1.1 Aerial Physical Interaction

One of the robotic fields in constant growth in the last decade is *aerial robotics*. According to [1], the definition of aerial robotics can be twofold: *(i) robotic flying machines*, putting the emphasis on the platform, or *(i) robotics that use flying machines*, putting the emphasis on the mission instead. In both cases, the main goal of aerial robotics is to study and conceive aerial systems that can perform work fully or partially autonomously. In the related literature, such robotic aerial platforms are often called *Unmanned Aerial Vehicles* (UAVs).

Although it is only recently that UAVs gained the interest of a very big and still increasing community, the study and design of such systems started already in the early 1900s. These vehicles were firstly used as prototypes to test new aircraft concepts before being produced and piloted by human pilots, decreasing the costs and the risks. The design of UAVs continued during the two World Wars for military purposes. However, the technology level was not enough to produce aerial robots able to autonomously navigate in a reliable way. It is only relatively recently, with the advent of lightweight and performing processors, accurate sensors and global navigation satellite systems, that aerial robots started to progressively gain better sensing and navigation capabilities. Although firstly employed in the military area, UAVs got a lot of interest from the civil area as well. Given the exponential appearance of new aerial vehicles and new applicative fields, the Economist compared the "drone boom" like the one happened to personal computers in the 1980s [2].

The motivation of the great popularity of UAVs mainly comes from the downscale of the size, weight, and cost of the sensing and computing technology. The latter made UAVs lighter, much more powerful and less expensive too. In turn, this allowed UAVs being accessible by a very wide community, both from the research and industrial areas. The low cost, the theoretically infinite workspace and the great versatility of these platforms allow employing them for several applications. In particular, they find their greatest use in dangerous and hazard environments, preventing

M. Tognon and A. Franchi, *Theory and Applications for Control of Aerial Robots in Physical Interaction Through Tethers*, Springer Tracts in Advanced Robotics 140,
https://doi.org/10.1007/978-3-030-48659-4_1

humans from getting harmed. Some examples of application where UAVs are nowadays employed are agriculture, construction, security, rescue, response to disasters, entertainment, photography and movie making, archeology and geographic mapping, wildlife monitoring/poaching, and many others can be mentioned. Other near-future interesting applications, currently under study, are personal and goods transportation (e.g., Volocopter[1] and Amazon,[2] respectively).

Several types of aerial vehicles are available in the market:

(1) *Rotary wings UAVs*, like multirotors, small-scale helicopters, and ducted fan;
(2) *Convertible UAVs*, like tail sitter aircrafts, that combine cruising flight and Vertical Takeoff and Landing (VTOL) capabilities [3];
(3) *Flapping wings UAVs*, inspired by the flight of birds, bats, and insects;
(4) *Fixed wings UAVs*, very popular for their long flight time.

According to the particular application, one could choose the vehicle that better fits the sought task, finding the best trade-off between flight endurance and maneuverability.

Particular attention is given to VTOL vehicles thanks to their high maneuverability and the ability to hover in place and to take off and land from/on confined spaces, without the need of a runway or other devices. These facts make VTOL vehicles applicable also in indoor and cluttered environments such as forests, industrial plants, and urban environments. A brief review of these type of vehicles is given in Sect. 3.2. Beyond the mentioned nice features of VTOL UAVs, they suffer from a major drawback. Standard VTOL vehicles, like collinear multirotors, can produce a total thrust force only along one fixed direction with respect to the body frame (they can also be called *unidirectional-thrust aerial vehicles*). This makes them *underactuated*. It means that we cannot fully control the vehicle state. In particular, one cannot control the attitude independently from the position. Starting from a hovering configuration (horizontal attitude), in order to move toward a certain direction the vehicle has firstly to rotate such that the thrust generates a horizontal acceleration toward the desired direction. This underactuation introduces several challenges for the stabilization of the system and the tracking of the desired trajectory. It also implies that an external disturbance cannot be immediately rejected. The platform has firstly to tilt. For these reasons, several works have been done to design controllers of increasing complexity to improve the performance of such vehicles, e.g., in [4–7]. Additionally, many state observers have been conceived to close the control loop to autonomously fly in different conditions and with different sensory setups. For more details on control, localization and motion planning methods for the navigation of VTOL UAVs, we refer the reader to the main surveys and books in the literature [8–12]

One can notice that in all the application mentioned so far, the robot is used as a simple remote sensor. The vehicle gathers data, e.g., with a camera, without interacting with the environment. Although the use of UAVs for applications concerning only the sensing of the environment is already an interesting and challenging topic,

---

[1] http://www.volocopter.com.

[2] http://www.amazon.com/Amazon-Prime-Air/b?ie=UTF8&node=8037720011.

it is actually limited with respect to the real potentiality of these aerial robots. The paramount scope of robots is to perform physical work, namely to act and interact with the environment exchanging forces. *Aerial Physical Interaction* (APhI) would lead to new very interesting applications. Some examples are assistance robotics in industrial or domestic environments, assembly and construction, decommissioning, inspection and maintenance by contact, removal of debris after natural disasters, delivery and transportation, stringing of power lines, and many others. Nowadays, these tasks are performed by human operators in very dangerous conditions, like on top of scaffolds or suspended by climbing ropes. The use of aerial robots would allow reducing the risk for the human operators and, at the same time, to reduce the cost associated with such operations. Given the relevance of the problem, many research labs and companies have been attracted to it. As a result, we can find many European projects with the scope of advancing in the aerial robotic field. In the following, we list some concluded and ongoing projects with corresponding goals:

- *ARCAS*[3]: conceive aerial robots for assembly and construction of structures;
- *AEROARMS*[4]: design and build UAVs with high manipulation capabilities for industrial inspection and maintenance;
- *AEROWORKS*[5]: provide heterogeneous and collaborative aerial robotic workers for inspection and maintenance tasks in infrastructure environments;
- *ARCOW*[6]: design aerial co-workers helping humans in manufacturing processes;
- *AEROBI*[7]: conceive aerial vehicles for in-depth structural inspection of concrete bridges;
- *AIROBOTS*[8]: design aerial robots for remote inspection by contact;
- *HYFLIERS*[9]: conceive a robot with hybrid air and ground mobility with a long-reach hyper-redundant manipulator.

For the aimed goals, aerial vehicles need new manipulation capabilities to safely and reliably interact with the environment. This opens the door to new challenges in aerial robotics. An aerial manipulator, being a floating body, has to *actively* react to interaction forces with the environment, that have to be carefully taken into account. Indeed they could eventually destabilize the system. This is different for a grounded manipulator which *passively* reacts to interaction forces thanks to the ground constraint. Furthermore, for grounded manipulators, we can usually directly and accurately control the torque that each motor applies to the corresponding joint. For an aerial vehicle, we instead control (in first approximation) the spinning velocity of a rotating propeller that, by the aerodynamic effects, produces a force. Due to the complexity of the aerodynamic effects and to disturbances, it is not easy to precisely

[3]http://www.arcas-project.eu/.

[4]https://aeroarms-project.eu/.

[5]http://www.aeroworks2020.eu/.

[6]http://www.euroc-project.eu/index.php?id=grvc-catec.

[7]http://www.aerobi.eu/.

[8]http://airobots.dei.unibo.it/.

[9]http://www.oulu.fi/hyfliers/.

control these forces. these actuation errors drastically impact the performance of the robot while interacting with the environment. Finally, in order to improve the dexterity and the manipulation capability of aerial robots, the latter are usually endowed with interactive tools such as grippers or articulated arms. The final aerial manipulator results to be a complex system characterized by a complicated and in general highly nonlinear dynamics. The latter has to be carefully considered because the couplings between the aerial robot and the interacting tool, if not properly addressed, can easily bring the system into instability. As a consequence, new control methods have to be conceived considering the full dynamics of the system, in order to preserve the stability during the interaction phases as well.

The most simple tool that one can use is a *rigid tool* rigidly attached to the robot. This allows exchanging forces with the environment, e.g., by pushing or sliding. Although the tool is very simple in itself, the underactuation of the vehicle makes the physical interaction very challenging. To address the problem, the works in [13–15] designed a hybrid force/position control. The tool can be then equipped with a *gripper* in order to allow pick and place operations [16]. To further increase the manipulation and the payload capabilities, several aerial robots endowed with a simple rigid link or a gripper can manipulate an object in a coordinated fashion, as a sort of "flying-hand" [17–19].

Another very used and still simple interaction tool is a *cable*. The use of a cable allows partially decoupling the rotational dynamics of the vehicle with respect to the one of the load. However, the control authority on the load positioning is reduced and a particular attention has to be given to undesired load oscillations that might destabilize the system. Several works addressed the problems from the control point of view proposing, for example, adaptive controllers [20, 21], a hierarchical controller [22], a flatness-based geometric controller [23] and even a reinforcement learning based approach [24]. Other works instead, addressed the problem from a motion planning point of view proposing algorithms that generate optimal trajectories that minimize the load swing [25, 26]. Also, in this case, the multi-robot approach can be beneficial to increase the payload of the system and the control authority on the load [27–32]. Furthermore, cables are not only used for the transportation of goods but also to tether aerial vehicles to fixed or moving platforms in order to enhance the flight stability during strong wind conditions or during dangerous maneuvers like takeoff and landing on moving vehicles [33, 34]. An introduction on the tethered aerial vehicle is provided in the following Sect. 1.2.

Finally, one can endow the aerial vehicle with one or even more *articulated arms*. The employ of a so-called *aerial manipulator* (AM) allows reaching high levels of dexterity. Depending on the number of degrees of freedom, an object can be locally manipulated independently from the motion of the platform. If the system is over-actuated, one can exploit the robot redundancy to achieve secondary tasks or to better compensate for external disturbances or tracking errors of the aerial vehicle. Nevertheless, the system results to be very complex and the underactuation of the vehicle makes its control even more complex. The easiest way to control such AM is with a *decentralized approach*. It consists of assuming the aerial vehicle and the robotic arm as two independent systems, considering the interaction forces as disturbances that have to be rejected. Indeed, the controller used for both subsystems is often a robust control [35, 36]. these methods can be also applied to robotic arms with kinematically controlled motors. However, they best perform only in quasi-static motions, i.e., when the couplings effects between the aerial vehicle and the articulated arm are practically negligible. As soon as the motion is more demanding in terms of accelerations, decentralized control methods fail, or in the best case show large tracking errors. In these cases is more advisable to use a *centralized control method* that considers the system as a unique entity. The centralized controllers proposed in the state of the art are strongly model-based and consider the full dynamics of the system [37–39]. If the kinematic and dynamic model is very well known, then centralized controllers can lead to very good performance. However, since they are strongly model based, as soon as there are some parameter uncertainties, the performance degrades. Furthermore, they often require torque controlled motors that are in general unfeasible for aerial manipulators due to the limited payload. A complete survey on the topic has also been recently published [40].

The previously mentioned examples address the aerial physical interaction problem using underactuated unidirectional-thrust vehicles. As already said this makes physical interaction tasks very challenging and prone to instability. However, a very recent and promising trend is to use *multidirectional-thrust aerial vehicles* instead [41]. As the name says, these vehicles can produce a thrust force in many directions with respect to the body frame. This means that they can independently control both position and orientation and can react to external disturbances almost instantaneously, when far from input saturation. These two great features make multidirectional-thrust aerial vehicles perfectly suited for physical interaction tasks since they are more robust to interaction forces and have more dexterity as well [42]. However, such benefit comes with the cost of a higher power consumption. In order to produce the thrust in several directions, the propellers are tilted or can be actively turned, toward different directions producing internal forces that waste energy. On the other hand, unidirectional-aerial vehicles are the most efficient in terms of energy. That is why it is still interesting to study aerial physical interaction by means of unidirectional-thrust vehicles.

Another important aspect of aerial physical interaction is related to motion planning. Even if we can control very well our robot, the trajectory for the execution of a certain task has to be carefully computed using motion planning techniques. To perform the task in a safe way, the planned trajectory must avoid obstacles and has to satisfy the intrinsic constraints of the considered robot. In particular, it has to be suitable for the dynamics of the system and its actuation limits. Classical motion planning methods rely on quasi-static assumptions and are based only on geometric and kinematic models of the system. Hence they are inadequate to achieve manipulation tasks involving physical interaction. In fact, when the robot is in contact with the environment and exchanges forces with it, the dynamics of the system cannot be neglected. This requires the use of a kinodynamic motion planning approach (e.g., [43]). However, kinodynamic planners developed so far are suitable only for simple systems, characterized by a small number of degrees of freedom and a relatively simple dynamic model, like car-like vehicles or quadrotors. Instead, in the context of aerial physical interaction, robots have usually a large number of degrees of freedom to increase the dexterity of the system. This, in turn, makes the motion planning problem very challenging, requiring the design of new kinodynamic motion planning methods. these have to cope with the nonlinear dynamics of the robot, its redundancy and the forces exchanged with the environment during manipulation tasks. Finally, the problem has to be solved very rapidly in order to use the planner online and to re-plan the trajectory in case of unforeseen events or moving obstacles. Some attempts to solve the motion planning problem for some specific cases can be found in [44, 45].

1.2 Tethered Aerial Vehicles

In the vast domain of UAVs, cables are not only used for single and cooperative transportation of goods. They are also used to tether the aerial vehicle to a ground station. Especially in the industrial sector, the link is mainly used as an umbilical device to provide power to the robot [46], and a high bandwidth communication channel with the base station. The possibility to power the robot directly from the ground station makes the aerial vehicle flight time theoretically infinite, overcoming one of the major limits of aerial robots. As a result, tethered aerial vehicles becomes suitable for many applications that require long operation time like monitoring [47], surveillance, aerial photography, communication reinforcement [48] and so on. The time flight provided by a single on-board battery would not be enough to fully accomplish the previous mentioned tasks. The great potentials of tethered aerial systems and their obtained big interest, is proven by the increasing number of private companies appeared in the market proposing tethered UAVs or power tether systems for standard commercial UAVs. Figure 1.1 gathers some of the many examples that one can find on-line.

(a) Courtesy of: Atlanta Instrumentation and Measurement, `www.aimatlanta.com`; and Guided Systems Technologies, `http://guidedsys.com`.

(b) Courtesy of: Elistair `www.elistair.com`.

(c) Elistair `www.elistair.com`. Copyright authorization in progress.

Fig. 1.1 Examples of companies proposing tethered aerial robots for long flight time operations

(a) Courtesy of: Apellix, `http://www.apellix.com/`

(b) Courtesy of: Aerones, `www.aerones.com`

Fig. 1.2 Examples of tethered aerial robots for cleaning applications

Another interesting use of the tether is to bring to the robot some sort of fluid for various type of applications, e.g., cleaning, painting or applying chemical products. Indeed, often there is the need of cleaning some part of a civil or industrial structure at high altitude, e.g. the windows of a skyscraper, the blades of a wind turbine, or the chimney of a refinery. Normally, these operations are conducted by human operators reaching the working spot by the use of climbing cords or by bulky and expensive scaffolds. Firstly, the use of an aerial robot in these applications would allow to perform the operation in an fully or semi autonomous mode reducing the risks for the human operators. Secondly, it would decrease the time and the costs related to the construction and deployment of scaffolds or climbing gears. However, due to the limited payload of standard aerial vehicles, it is practically unfeasible to carry on-board all the tools needed for these kind of tasks, e.g., a spying tool and a tank of detergent liquid. On the other hand, the tether could be made such that to provide to the robot not only the power to fly indefinitely, but the cleaning liquid as well. Figure 1.2 shows the tethered aerial vehicles proposed by two companies for the cleaning of a facade of an industrial structure, and the blades of a wind turbine, respectively.

In the previous mentioned cases, the cable is slack, i.e., there is not tension along the link. Therefore, except for its weight and inertia, the cable does not influence the motion of the aerial vehicle. In these cases, complex control strategy are not really needed and a standard position controller (or a tele-operation framework) can be used to perform the sought task.

The case in which the cable is taut is definitely more interesting from a scientific point of view. In this case there is a clear physical interaction between the aerial vehicle, the link itself, and the other end of the link. A taut cable can provide advantages that go beyond the ones already mentioned, such as: (i) improved flight stability and reliability, especially during dangerous maneuver or in the presence of strong wind [33], (ii) physical interaction with a ground object and (iii) stabilization with a minimal set of sensors, even in a GPS-denied environment [34, 49]. Examples of

(a) EC-SAFEMOBIL Copyright authorization in progress.

(b) FotoKite. Copyright authorization in progress.

Fig. 1.3 Examples of applications of tethered aerial vehicles when the cable is taut. In particular, starting from the left image: landing/takeoff on/from a moving platform and indoor inspection (a) Courtesy of: EC-SAFEMOBIL [53–55] (b) Courtesy of: FotoKite www.fotokite.com

application fields related to this kind of robotic systems exploiting the tautness of the cable are landing/taking-off from/on moving or sloped platforms [33, 50–52], inspection in GPS-denied environments, human-robot interaction [34] and stringing of power transmission lines (see Fig. 1.3 for some examples).

Notice that, since the link is taut, the dynamics of the aerial vehicles changes. Indeed the interaction force consisting in the internal force along the link has to be considered. Control and estimation for an aerial system that is connected by a taut cable to the ground is not an easy task. In fact, standard flight-control and estimation methods either cannot be applied straightforwardly to this case or, if applicable, provide only sub-optimal performance, because they do not exploit the full dynamics and capabilities available to the new interconnected system. Therefore, the only way to cope with the difficulties of the new robotic system and to exploit at best its capabilities is to design new control and estimation methods that consider the new system as whole. However this is hard to accomplish, due to the nonlinear dynamics and the dynamic coupling between the aerial vehicle and the link.

For a complete understanding of these type of robotic systems, this book aims at providing a deep and thorough theoretical analysis which is the basis for solving practical problems related to real applications. In particular, we shall consider the most generic tethered aerial system, i.e., a generic unidirectional-thrust aerial vehicle flying in the 3D space and tethered to a freely moving platform by a generic link (not only by a cable), together with a link actuator able to change its length. For this system, we shall investigate its dynamics and its intrinsic properties, such as the *differential flatness*. This property is very useful both for control and motion planning. Indeed the analysis of such property allows understanding which are the outputs, called *flat outputs*, that can be independently controlled and which is their required degree of smoothness. Furthermore, it provides the tools to analytically compute the nominal state and input required to track a desired trajectory of the flat output. This turns out to be very helpful in the motion planning phase to simplify the planning method and to check for the feasibility of the desired trajectory. These results will be the base to design different types of *controllers for tracking* the outputs of interest. To close the control loop, the full state of the system is required. Practically, having a direct measurement of the state is often unfeasible. In this work we shall then considered the problem of closing the control loop with a minimal set of sensors, investigating the *observability*, and designing a *global nonlinear observer*.

Finally, we will try to apply the presented theoretical results to a real application problem. In particular we shall consider the practical problem of *takeoff and landing from/on a sloped surface*. For a standard unidirectional-thrust vehicle in a free-flight configuration, this is a very challenging problem. On the other hand, we will theoretically and experimentally show that the use of the tether makes the execution of these maneuvers much safer, reliable and robust to tracking errors and parameters uncertainties.

1.3 Organization of the Book

In this section, we provide a reader's guide describing the organization of the book and summarizing the content of each chapter.

The first three chapters provide the preliminaries to better contextualize this work in the field of aerial physical interaction, and the mathematical methodologies to better understand the core part. In particular,

Chapter 2 recalls in a synthetic way the mathematical methodologies used as background for the theoretical analysis of tethered aerial vehicles. In particular, we revise the two most used modeling methods, i.e., Lagrangian and Newton-Euler formalisms, the differential flatness property, the dynamic feedback linearization control, and the nonlinear high gain observer.

Chapter 3 provides the models of the subsystems, actuators, and sensors composing the studied generic tethered aerial vehicle. We provide the generic model of an unidirectional-thrust aerial vehicle in free-flight, the model for its propellers and onboard sensors. A model of a generic link and of an encoder are also provided.

Chapters from 4 to 6 represent the main body of the book, namely the complete and exhaustive study of aerial tethered vehicles. This analysis starts with the modeling of the system and passes through the characterization of its dynamic properties such as the differential flatness, controllability, and observability using a minimal sensory setup. The results of an experimental and simulation campaign are presented to validate the proposed methods. These are also used as a base for solving the more applicative problem of landing and takeoff on/from a sloped surface. In particular, we shall show that the use of a tether makes the execution of such dangerous maneuvers much more safe, reliable and robust to model uncertainties and tracking errors. Finally, a multi-robot extension is considered for which we performed a thorough theoretical analysis similar to the one for the single-tethered case. This part gathers the work of several articles and some unpublished results as well. For the sake of homogeneity, completeness, and clarity, in this book we present a complete dissertation of the topic based on a very detailed reworking of the content of the several publications.

Chapter 4 contains the complete and thorough theoretical study of a single tethered aerial vehicle. We provide the model of a generic system. For such a system we investigate the differential flatness and which are the flat outputs. For the latter, we design a hierarchical controller and another controller based on dynamic feedback linearization for the tracking of any desired trajectory. Finally, we investigate the problem of closing the control loop with a minimal sensory setup.
Chapter 5 presents all the results obtained from an extensive experimental and numerical campaign apt to validate the proposed methods.
Chapter 6 shows the study of the challenging and application-oriented problem of landing and takeoff on/from a sloped surface. For this problem, we theoretically and experimentally show that the use of the tether is advisable, when possible. Indeed, it allows executing these maneuvers in a much more robust and reliable way.
Chapter 7 analyzes an interesting multi-robot extension of the single tethered system. This system is similar to a 2-links planar manipulator where the actuators are aerial vehicles connected in a chain-like configuration. For this system, similarly to the single-robot system, we analyze the differential flatness, the controllability by dynamic feedback linearization and the observability using a minimal set of sensors.

The book concludes with Chap. 8:

Chapter 8 provides a global overview of the book, together with some discussions about the obtained results. Potential future applications and extensions of this work are considered as well.

References

1. Feron, E., Johnson, E.N.: Aerial robotics. In: Springer Handbook of Robotics, pp. 1009–1029. Springer (2008)
2. Economist, T.: Welcome to the Drone Age, Sept (2015). [Online]
3. Morin, P.: Modeling and control of convertible micro air vehicles. In: 2015 10th International Workshop on Robot Motion and Control (RoMoCo), pp. 188–198. IEEE (2015)
4. Hua, M.-D., Hamel, T., Morin, P., Samson, C.: Control of VTOL vehicles with thrust-tilting augmentation. Automatica **52**, 1–7 (2015)
5. Plinval, H., Morin, P., Mouyon, P., Hamel, T.: Visual servoing for underactuated VTOL UAVs: a linear, homography-based framework. Int. J. Robust Nonlinear Control **24**(16), 2285–2308 (2014)

6. Faessler, M., Franchi, A., Scaramuzza, D.: Differential flatness of quadrotor dynamics subject to rotor drag for accurate tracking of high-speed trajectories. IEEE Robot. Autom. Lett. **3**(2), 620–626 (2018)
7. Lee, T., Leoky, M., McClamroch, N.H.: Geometric tracking control of a quadrotor UAV on SE(3). In: 49th IEEE Conference on Decision and Control, pp. 5420–5425. Atlanta, GA, Dec (2010)
8. Ollero, A., Merino, L.: Control and perception techniques for aerial robotics. Annu. Rev. Control **28**(2), 167–178 (2004)
9. Valavanis, K.P., Vachtsevanos, G.J.: Future of unmanned aviation. In: Handbook of Unmanned Aerial Vehicles, pp. 2993–3009. Springer (2015)
10. Mahony, R., Kumar, V., Corke, P.: Multirotor aerial vehicles: modeling, estimation, and control of quadrotor. IEEE Robot. Autom. Mag. **19**(3), 20–32 (2012)
11. Hua, M.-D., Hamel, T., Morin, P., Samson, C.: Introduction to feedback control of underactuated VTOL vehicles: a review of basic control design ideas and principles. IEEE Control Syst. Mag. **33**(1), 61–75 (2013)
12. Hua, M.-D., Hamel, T., Morin, P., Samson, C.: A control approach for thrust-propelled underactuated vehicles and its application to VTOL drones. IEEE Trans. Autom. Control **54**(8), 1837–1853 (2009)
13. Nguyen, H., Lee, D.: Hybrid force/motion control and internal dynamics of quadrotors for tool operation. In: 2013 IEEE/RSJ International Conference on Intelligent Robots and Systems, pp. 3458–3464, Tokyo, Japan, Nov (2013)
14. Bartelds, T., Capra, A., Hamaza, S., Stramigioli, S., Fumagalli, M.: Compliant aerial manipulators: toward a new generation of aerial robotic workers. IEEE Robot. Autom. Lett. **1**(1), 477–483 (2016)
15. Gioioso, G., Mohammadi, M., Franchi, A., Prattichizzo, D.: A force-based bilateral teleoperation framework for aerial robots in contact with the environment. In: 2015 IEEE International Conference on Robotics and Automation, pp. 318–324, Seattle, WA, May (2015)
16. Mellinger, D., Lindsey, Q., Shomin, M., Kumar, V.: Design, modeling, estimation and control for aerial grasping and manipulation. In: 2011 IEEE/RSJ International Conference on Intelligent Robots and Systems, pp. 2668–2673, San Francisco, CA, Sept (2011)
17. Mellinger, D., Shomin, M., Michael, N., Kumar, V.: Cooperative grasping and transport using multiple quadrotors. In: 10th International Symposium on Distributed Autonomous Robotic Systems, pp. 545–558, Lausanne, Switzerland, Nov (2010)
18. Gioioso, G., Salvietti, G., Franchi, A., Malvezzi, M., Scheggi, S., Meli, L., Ryll, M., Bülthoff, H.H., Prattichizzo, D.: The flying hand: a teleoperation framework for cooperative aerial grasping and transportation. In: Automatica.it 2014, Convegno Annuale dei Docenti e Ricercatori Italiani in Automatica, Bergamo, Italy, Sept (2014)
19. Staub, N., Mohammadi, M., Bicego, D., Prattichizzo, D., Franchi, A.: Towards robotic MAGMaS: multiple aerial-ground manipulator systems. In: 2017 IEEE International Conference on Robotics and Automation, Singapore, May (2017)
20. Dai, S., Lee, T., Bernstein, D.S.: Adaptive control of a quadrotor UAV transporting a cable-suspended load with unknown mass. In: 2014 IEEE 53rd Annual Conference on Decision and Control (CDC), pp. 6149–6154. IEEE (2014)
21. Palunko, I., Cruz, P., Fierro, R.: Agile load transportation: safe and efficient load manipulation with aerial robots. IEEE Robot. Autom. Mag. **19**(3), 69–79 (2012)
22. Bernard, M., Kondak, K., Maza, I., Ollero, A.: Autonomous transportation and deployment with aerial robots for search and rescue missions. J. Field Robot. **28**(6), 914–931 (2011)
23. Sreenath, K., Lee, T., Kumar, V.: Geometric control and differential flatness of a quadrotor UAV with a cable-suspended load. In: 2013 IEEE 52nd Annual Conference on Decision and Control (CDC), pp. 2269–2274. IEEE (2013)
24. Palunko, I., Faust, A., Cruz, P., Tapia, L., Fierro, R.: A reinforcement learning approach towards autonomous suspended load manipulation using aerial robots. In: 2013 IEEE International Conference on Robotics and Automation (ICRA), pp. 4896–4901. IEEE (2013)

25. Sreenath, K., Michael, N., Kumar, V.: Trajectory generation and control of a quadrotor with a cable-suspended load-a differentially-flat hybrid system. In: 2013 IEEE International Conference on Robotics and Automation (ICRA), pp. 4888–4895. IEEE (2013)
26. Foehn, P., Falanga, D., Kuppuswamy, N., Tedrake, R., Scaramuzza, D.: Fast trajectory optimization for agile quadrotor maneuvers with a cable-suspended payload. In: Robotics: Science and Systems, pp. 1–10 (2017)
27. Sreenath, K., Kumar, V.: Dynamics, control and planning for cooperative manipulation of payloads suspended by cables from multiple quadrotor robots. In: Robotics: Science and Systems, Berlin, Germany, June (2013)
28. Manubens, M., Devaurs, D., Ros, L., Cortés, J.: Motion planning for 6-D manipulation with aerial towed-cable systems. In: Robotics: Science and Systems, p. 2013. Germany, May, Berlin (2013)
29. Maza, I., Kondak, K., Bernard, M., Ollero, A.: Multi-UAV cooperation and control for load transportation and deployment. J. Intell. Robot. Syst. **57**(1–4), 417–449 (2010)
30. Gabellieri, C., Tognon, M., Sanalitro, D., Palottino, L., Franchi, A.: A study on force-based collaboration in swarms. Swarm Intell. **14**, 57–82 (2020)
31. Tognon, M., Gabellieri, C., Pallottino, L., Franchi, A.: Aerial co-manipulation with cables: the role of internal force for equilibria, stability, and passivity. IEEE Robot. Autom. Lett. Spec. Issue Aerial Manip. **3**(3), 2577–2583 (2018)
32. Sanalitro, D., Savino, H.J., Tognon, M., Cortés, J., Franchi, A.: Full-pose manipulation control of a cable-suspended load with multiple UAVs under uncertainties. Under review in IEEE Robotics and Automation Letters (2020)
33. Sandino, L.A., Bejar, M., Kondak, K., Ollero, A.: Advances in modeling and control of tethered unmanned helicopters to enhance hovering performance. J. Intell. Robot. Syst. **73**(1–4), 3–18 (2014)
34. Lupashin, S., D'Andrea, R.: Stabilization of a flying vehicle on a taut tether using inertial sensing. In: 2013 IEEE/RSJ International Conference on Intelligent Robots and Systems, pp. 2432–2438, Tokyo, Japan, Nov (2013)
35. Siciliano, B., Sciavicco, L., Villani, L., Oriolo, G.: Robotics: Modelling. Springer, Planning and Control (2009)
36. Ruggiero, F., Trujillo, M.A., Cano, R., Ascorbe, H., Viguria, A., Peréz, C., Lippiello, V., Ollero, A., Siciliano, B.: A multilayer control for multirotor UAVs equipped with a servo robot arm. In: 2015 IEEE International Conference on Robotics and Automation, pp. 4014–4020, Seattle, WA, May (2015)
37. Yang, H., Lee, D.J.: Dynamics and control of quadrotor with robotic manipulator. In: 2014 IEEE International Conference on Robotics and Automation, pp. 5544–5549, Hong Kong, China, May (2014)
38. Yüksel, B., Staub, N., Franchi, A.: Aerial robots with rigid/elastic-joint arms: single-joint controllability study and preliminary experiments. In: 2016 IEEE/RSJ International Conference on Intelligent Robots and Systems, pp. 1667–1672, Daejeon, South Korea, Oct (2016)
39. Mersha, A.Y., Stramigioli, S., Carloni, R.: Exploiting the dynamics of a robotic manipulator for control of UAVs. In: 2014 IEEE International Conference on Robotics and Automation (ICRA), pp. 1741–1746. IEEE (2014)
40. Ruggiero, F., Lippiello, V., Ollero, A.: Aerial manipulation: a literature review. IEEE Robot. Autom. Lett. **3**(3), 1957–1964 (2018)
41. Rajappa, S., Ryll, M., Bülthoff, H.H., Franchi, A.: Modeling, control and design optimization for a fully-actuated hexarotor aerial vehicle with tilted propellers. In: 2015 IEEE International Conference on Robotics and Automation, pp. 4006–4013, Seattle, WA, May (2015)
42. Ryll, M., Muscio, G., Pierri, F., Cataldi, E., Antonelli, G., Caccavale, F., Franchi, A.: 6D physical interaction with a fully actuated aerial robot. In: 2017 IEEE International Conference on Robotics and Automation, pp. 5190–5195, Singapore, May (2017)
43. Boeuf, A., Cortés, J., Alami, R., Siméon, T.: Enhancing sampling-based kinodynamic motion planning for quadrotors. In: 2015 IEEE/RSJ International Conference on Intelligent Robots and Systems, pp. 2447–2452, Hamburg, Germany, Sept (2015)

44. Lee, H., Kim, H., Kim, H.J.: Planning and control for collision-free cooperative aerial transportation. IEEE Trans. Autom. Sci, Eng (2016)
45. Fink, J., Michael, N., Kim, S., Kumar, V.: Planning and control for cooperative manipulation and transportation with aerial robots. Int. J. Robot. Res. **30**(3), 324–334 (2011)
46. Choi, S.Y., Choi, B.H., Jeong, S.Y., Gu, B.W., Yoo, S.J., Rim, C.T.: Tethered aerial robots using contactless power systems for extended mission time and range. In: Energy Conversion Congress and Exposition (ECCE), 2014 IEEE, pp. 912–916. IEEE (2014)
47. Muttin, F.: Umbilical deployment modeling for tethered UAV detecting oil pollution from ship. Appl. Ocean Res. **33**(4), 332–343 (2011)
48. Pinkney, M.F., Hampel, D., DiPierro, S.: Unmanned aerial vehicle (UAV) communications relay. In: Military Communications Conference, 1996, vol. 1, pp. 47–51, Oct (1996)
49. Sandino, L., Santamaria, D., Bejar, M., Viguria, A., Kondak, K., Ollero, A.: Tether-guided landing of unmanned helicopters without GPS sensors. In: 2014 IEEE International Conference on Robotics and Automation, pp. 3096–3101, Hong Kong, China, May (2014)
50. Tognon, M., Dash, S.S., Franchi, A.: Observer-based control of position and tension for an aerial robot tethered to a moving platform. IEEE Robot. Autom. Lett. **1**(2), 732–737 (2016)
51. Tognon, M., Testa, A., Rossi, E., Franchi, A.: Takeoff and landing on slopes via inclined hovering with a tethered aerial robot. In: 2016 IEEE/RSJ International Conference on Intelligent Robots and Systems, pp. 1702–1707, Daejeon, South Korea, Oct (2016)
52. Oh, S.-R., Pathak, K., Agrawal, S.K., Pota, H.R., Garrett, M.: Approaches for a tether-guided landing of an autonomous helicopter. IEEE Trans. Robot. **22**(3), 536–544 (2006)
53. EC-SAFEMOBIL. EU Coll. Proj. FP7-ICT 288082. http://www.ec-safemobil-project.eu/
54. Alarcón, F., García, M., Maza, I., Viguria, A., Ollero, A.: A Precise and GNSS-Free Landing System on Moving Platforms for Rotary-Wing UAVs. SensS. **19**(4), 886 (2019)
55. Sandino, Luis, A., Santamaria, D., Bejar, M., Kondak, K., Viguria, A., Ollero, A.: First experimental results on enhancing hovering performance of unmanned helicopters by using a tethered setup. Robot. Auton. Syst. **79**(4), 147–155 (2016)

Chapter 2
Theoretical Background

2.1 Modeling

The major two methodologies normally employed to compute the dynamic model of a mechanical system are the *Lagrangian* and the *Newton-Euler* formalisms. The two methods are equivalent and obviously lead to the same outcome, but the practical procedure is quite different. Furthermore, they could give different insights about the system and its properties.

The first, the Lagrangian formalism, is a systematic and elegant approach to derive the analytical dynamic equations describing the model of the system, independently from the reference frame. In particular, choosing a proper set of generalized coordinates and simply computing the kinematics and potential energies, the Lagrangian formalism allows to compute the dynamic equations, naturally including system constraints and reaction forces. Nevertheless, notice that it becomes unpractical for complex system with many degrees of freedom.

On the other hand, the Newton-Euler method is an efficient and recursive method, especially suited for manipulators with an open kinematic chain and complex systems. It treats each joint of a robot as an independent part, and then computes the coupling between them using the so called *forward-backward* recursive algorithm. However, a particular attention has to be taken for constrained systems. Indeed one has to explicitly consider reaction forces related to system constraints.

In the following we recall the basis of the two methods, mostly from a practical point of view, and the particular remarks and considerations made during this thesis. Fore more details we refer the interested reader to [2–5].

2.1.1 *Lagrange Formalism*

The fist step consists on choosing a set of independent coordinates $\mathbf{q} = [q_1 \ \ldots \ q_n]^\top \in \mathbb{R}^n$, called *generalized coordinates*. Those fully describe the configuration of the

M. Tognon and A. Franchi, *Theory and Applications for Control of Aerial Robots in Physical Interaction Through Tethers*, Springer Tracts in Advanced Robotics 140,
https://doi.org/10.1007/978-3-030-48659-4_2

system and its $n \in \mathbb{N}_{>0}$ degrees of freedom. Accordingly to the chosen generalized coordinates, we can then compute the *generalized forces* acting on the system. Consider a set of forces $\mathbf{f} = [\mathbf{f}_1^\top \ \dots \ \mathbf{f}_m^\top]^\top \in \mathbb{R}^{3m}$, where the generic force $\mathbf{f}_i \in \mathbb{R}^3$ is applied on the system at point $\mathbf{r}_i \in \mathbb{R}^3$, with $i = 1, \dots, m$ and $m \in \mathbb{N}_{\geq 0}$. We can then compute the generalized force $\xi_j(\mathbf{f}, \mathbf{q}) \in \mathbb{R}$ w.r.t. the jth generalized coordinate q_j as:

$$\xi_j(\mathbf{f}, \mathbf{q}) = \sum_{i=1}^{m} \mathbf{f}_i^\top \frac{\partial \mathbf{r}_i}{\partial q_j}, \quad j = 1, \dots, n. \tag{2.1}$$

We can now define the *Lagrangian* function, $\mathcal{L}(\mathbf{q}, \dot{\mathbf{q}})$, equal to the difference of total kinetic energy, $\mathcal{K}(\mathbf{q}, \dot{\mathbf{q}})$, and potential energy, $\mathcal{U}(\mathbf{q}, \dot{\mathbf{q}})$, i.e., $\mathcal{L}(\mathbf{q}, \dot{\mathbf{q}}) = \mathcal{K}(\mathbf{q}, \dot{\mathbf{q}}) - \mathcal{U}(\mathbf{q}, \dot{\mathbf{q}})$. Finally, the equation of motions of the system are given by the following Lagrange equations:

$$\frac{d}{dt}\frac{\partial \mathcal{L}(\mathbf{q}, \dot{\mathbf{q}})}{\partial \dot{q}_j} - \frac{\partial \mathcal{L}(\mathbf{q}, \dot{\mathbf{q}})}{\partial q_j} = \xi_j(\mathbf{f}, \mathbf{q}), \quad j = 1, \dots, n. \tag{2.2}$$

For the type of mechanical systems under exam, the potential energy usually corresponds to the sole gravitational potential energy, and the kinematic energy can be computed as a quadratic form, $\mathcal{K}(\mathbf{q}, \dot{\mathbf{q}}) = \frac{1}{2}\dot{\mathbf{q}}^\top \mathbf{M}(\mathbf{q})\dot{\mathbf{q}}$, where $\mathbf{M}(\mathbf{q}) \in \mathbb{R}^{n \times n}$ is the inertia matrix of the system. The equations of motion in (2.2) can be then rewritten in the more usual form:

$$\mathbf{M}(\mathbf{q})\ddot{\mathbf{q}} + \mathbf{C}(\mathbf{q}, \dot{\mathbf{q}})\dot{\mathbf{q}} + \mathbf{g}(\mathbf{q}) = \boldsymbol{\xi}(\mathbf{f}, \mathbf{q}), \tag{2.3}$$

where $\mathbf{C}(\mathbf{q}, \dot{\mathbf{q}})\dot{\mathbf{q}}$ contains the centrifugal and Coriolis terms, while $\mathbf{g}(\mathbf{q})$ contains the gravitational terms, and $\boldsymbol{\xi}(\mathbf{f}, \mathbf{q}) = [\xi_1(\mathbf{f}, \mathbf{q}) \ \dots \ \xi_n(\mathbf{f}, \mathbf{q})]^\top \in \mathbb{R}^n$.

Remark The *inverse dynamics* problem consists into computing the generalized forces $\boldsymbol{\xi}(\mathbf{f}, \mathbf{q})$ given a certain motion expressed in terms of $\ddot{\mathbf{q}}$, $\dot{\mathbf{q}}$ and $\mathbf{q}$. Considering the generalized forces as inputs and the motion as output, this problem is equivalent to the control problem, i.e., compute certain inputs to obtain certain desired outputs. Given the analytic expression of the dynamic model (2.3), the Lagrangian formalism is often used to solve the *inverse dynamics* problem, and thus the control problem. □

2.1.2 Newton-Euler Formalism

The Newton-Euler formalism is based on two recursive steps: *(i)* forward recursion, and *(ii)* backward recursion.

The first *forward recursion* is done to propagate the links velocities and accelerations from the first link to the final one. The translational and rotational velocities and acceleration of the ith link are computed based on the one of the previous $(i-1)$th link and on the ith joint, according to its type (either prismatic or revolute). The method is repeated for all the links starting from the base link, of which we know velocities and accelerations, up to the last one.

The second *backward recursion* propagates forces and moments from the last link to the first one. Knowing the force and moment applied to the $(i+1)$th link, we compute the one applied to the ith link resolving the Newton-Euler equations. Defining $\mathbf{f}_i \in \mathbb{R}^3$ and $\boldsymbol{\tau}_i \in \mathbb{R}^3$ the force and moment acting on the ith link at position $\mathbf{r}_i \in \mathbb{R}^3$ (analogously for the $(i+1)$th link), we have to solve the balance equations of forces and moments at the i-link w.r.t. the ith link frame:

$$\mathbf{f}_i = \mathbf{f}_{i+1} + m_i \mathbf{a}_i + m_i \mathbf{g}_i \tag{2.4a}$$

$$\boldsymbol{\tau}_i = \boldsymbol{\tau}_i - \mathbf{f}_i \times \mathbf{r}_i + \mathbf{f}_{i+1} \times \mathbf{r}_{i+1} + \mathbf{J}_i \dot{\boldsymbol{\omega}}_i + \boldsymbol{\omega}_i \times \mathbf{J}_i \boldsymbol{\omega}_i, \tag{2.4b}$$

where $m_i \in \mathbb{R}_{>0}$ and $\mathbf{J}_i \in \mathbb{R}^{3\times 3}_{>0}$ are the mass and inertia[1] of the ith link, $\mathbf{a}_i \in \mathbb{R}^3$ is its linear acceleration, $\boldsymbol{\omega}_i \in \mathbb{R}^3$ and $\dot{\boldsymbol{\omega}}_i \in \mathbb{R}^3$ are its angular velocity and acceleration, respectively, and $\mathbf{g}_i \in \mathbb{R}^3$ is the gravity vector. Notice that all the previous quantities are defined w.r.t. the ith link frame. The method is repeated for all the links starting from the final one, whose external forces and moments are known, back to the first one.

Finally one could retrieve a closed form dynamic model, like the one in (2.3), resolving all together the forward and backward equations. However, doing it analytically might not be an easy task. We skip the detailed equations because of their complexity. Nevertheless, we refer the interested reader to the well known books [2–5].

2.1.3 *Rigid Body Dynamics*

In view of the fact that an aerial vehicle is often modeled as a rigid body, it is convenient here to review the dynamic model of such a basic element. A free rigid body, i.e., not subjected to constraints, has six degrees of freedom: three translational and three rotational. Let us assign an inertial *word frame*, $\mathcal{F}_W$ with arbitrary center O_W and axes $\{\mathbf{x}_W, \mathbf{y}_W, \mathbf{z}_W\}$, and a *body frame*, $\mathcal{F}_B$, rigidly attached to the object, with center O_B centered on the body center of mass (CoM), and axes $\{\mathbf{x}_B, \mathbf{y}_B, \mathbf{z}_B\}$. It is useful to notice here that $\mathbf{x}_W^W = \mathbf{e}_1 = [1\ 0\ 0]^\top$, $\mathbf{y}_W^W = \mathbf{e}_2 = [0\ 1\ 0]^\top$ and[2] $\mathbf{z}_W^W = \mathbf{e}_3 = [0\ 0\ 1]^\top$. The three translational degrees of freedom are described by the position of O_B with respect to $\mathcal{F}_W$, in turn described by the vector[3] $\mathbf{p}_B^W \in \mathbb{R}^3$. The description of the remaining

[1]The notation $\mathbb{R}^{n\times n}_{>0}$ denotes the set of positive-definite real matrices, i.e., $\mathbb{R}^{n\times n}_{>0} = \{\mathbf{A} \in \mathbb{R}^{n\times n} \mid \mathbf{x}^\top \mathbf{A} \mathbf{x} > 0\ \forall\, \mathbf{x} \in \mathbb{R}^n\}$.

[2]More in general, $\mathbf{e}_i \in \mathbb{R}^3$ is the canonical vector with 1 in position ith and zero otherwise.

[3]In this thesis, the superscript is used to indicate the frame of references. When not present, $\mathcal{F}_W$ has to be intended as the reference frame, if not otherwise specified.

three orientation degrees of freedom is a bit more delicate because there are several possible representations [3, 4, 6]. The most popular and used are:

- The *exponential coordinates* are a minimal three-parameter representation of rotations which define an axis of rotation and the corresponding angle of rotation. However, combinations of rotations is not straightforward and the axis of rotation is undetermined when the angle of rotation goes to zero.
- The *Euler-angles* is another minimal three-parameter representation of rotations. It is also very intuitive, since it is based on three successive rotations about the main axes of the body frame. One of the most popular convention in the aeronautic field consists in successive rotations along the moving axes $\mathbf{z}_B$, $\mathbf{y}_B$ and $\mathbf{x}_B$ (in this order) about the angles ψ, θ and ϕ (*Yaw-Pitch-Roll*) respectively.[4] However, this representation has a singularity. To avoid singularities at the control level, we will use this convention only to represent rotations in plots.
- The *rotation matrix*,[5] $\mathbf{R}_B^W \in \mathsf{SO}(3)$, unequivocally describes the rotation of $\mathcal{F}_B$ w.r.t. $\mathcal{F}_W$. Although this representation has no singularities, it is actually redundant since nine elements describe only three degrees of freedom. Nevertheless, it eases the operations to rotate vectors and to combine rotations. These facts together with the absence of singularities make this representation the preferable for the design of controllers for aerial vehicles. This is why, in this thesis, we will always describes rotations by rotation matrices.
- The *quaternions* represent rotations by a normalized four-dimensional vector, i.e., four variables subjected to one constraint. In this way, the quaternion parametrization does not have singularities. This parametrization is also very popular for it efficiency in terms of computational cost. However, in this thesis we still prefer rotation matrices for their simplicity. This will clearly appear in Chap. 4.

Choosing $\mathbf{p}_B^W$ and $\mathbf{R}_B^W$ to describe the rigid body configuration, we can write the dynamics as in (2.4), using the Newton-Euler approach:

$$m\ddot{\mathbf{p}}_B^W = -mg\mathbf{e}_3 + \mathbf{f} \tag{2.5a}$$

$$\mathbf{J}\dot{\boldsymbol{\omega}}_B^B = -\boldsymbol{\omega}_B^B \times \mathbf{J}\boldsymbol{\omega}_B^B + \boldsymbol{\tau}, \tag{2.5b}$$

where $m \in \mathbb{R}_{>0}$ and $\mathbf{J} \in \mathbb{R}_{>0}^{3\times 3}$ are the mass and inertia of the rigid body w.r.t. $\mathcal{F}_B$, $\ddot{\mathbf{p}}_B^W \in \mathbb{R}^3$ is its linear acceleration, $\boldsymbol{\omega}_B^B \in \mathbb{R}^3$ and $\dot{\boldsymbol{\omega}}_B^B \in \mathbb{R}^3$ are its angular velocity and acceleration w.r.t. $\mathcal{F}_W$ expressed in $\mathcal{F}_B$, respectively, $g \approx 9.81$ is the gravitational constant, $\mathbf{f} \in \mathbb{R}^3$ and $\boldsymbol{\tau} \in \mathbb{R}^3$ are the sum of forces and moments applied to the body CoM, respectively. Furthermore we recall the differential kinematic relation $\dot{\mathbf{R}}_B^W = \mathbf{R}_B^W \boldsymbol{\Omega}_B^B$, where $\boldsymbol{\Omega}_\star$ is the skew symmetric matrix associated to $\boldsymbol{\omega}_\star$.

---

[4]Notice that this representation is equivalent to the classical Roll-Pitch-Yaw representation. The latter consists in successive rotations along the fixed axes $\mathbf{x}_B$, $\mathbf{y}_B$ and $\mathbf{z}_B$ (in this order) about the angles ϕ, θ and ψ respectively.

[5]$\mathsf{SO}(3) = \{\mathbf{R} \in \mathbb{R}^{3\times 3} \mid \mathbf{R}^\top\mathbf{R} = \mathbf{I}_3\}$ where $\mathbf{I}_n$ is the identity matrix of dimension n. $\mathsf{SO}(3)$ is also called *special orthogonal group*.

As seen before, when it comes to model a floating vehicle, the use of rotation matrix representation and Newton-Euler method is really convenient. The Lagrange method would have instead required the use of minimal representation for the orientation. Nevertheless, as previously said, Newton-Euler method is not favorable in the presence of constraints and reaction forces. Therefore, the approach employed in this thesis tries to exploit the good features of both Lagrangian and Newton-Euler methods. In particular, in order to model a tethered aerial vehicle, in Sect. 4.3 we firstly use the Lagrangian formalism to identify the most convenient generalized coordinates describing the translational dynamics of the vehicle subjected to the constraint given by the link. We instead used a rotation matrix for the description of the attitude. Afterwards, we applied the Newton-Euler method to retrieve the dynamics of the system and the analytical expression of the internal force. Since one of the control objectives is the precise control of the internal force, the analytical expression will be useful to design a tracking controller based on dynamic feedback linearization.

2.2 Differential Flatness

For the analysis of nonlinear dynamic systems, one important property to verify is the *differential flatness*. This property was firstly introduced by Michel Fliess in the late 1980s, and then exploited in many other works for the control of nonlinear systems [7–9]. The formal definition of a differentially flat systems follows:

Definition A system $\dot{\mathbf{x}} = \mathbf{f}(\mathbf{x}, \mathbf{u})$ with state vector $\mathbf{x} \in \mathbb{R}^n$ and input vector $\mathbf{u} \in \mathbb{R}^m$, where $\mathbf{f}$ is a smooth vector field, is *differentially flat* if it exists an output vector $\mathbf{y} \in \mathbb{R}^m$, called *flat output*, in the form:

$$\mathbf{y} = \mathbf{h}(\mathbf{x}, \mathbf{u}, \dot{\mathbf{u}}, \ldots, \mathbf{u}^{(q)}) \tag{2.6}$$

such that

$$\mathbf{x} = \mathbf{g}_{\mathbf{x}}(\mathbf{y}, \dot{\mathbf{y}}, \ldots, \mathbf{y}^{(r)}) \tag{2.7}$$

$$\mathbf{u} = \mathbf{g}_{\mathbf{u}}(\mathbf{y}, \dot{\mathbf{y}}, \ldots, \mathbf{y}^{(r)}) \tag{2.8}$$

where[6] $\mathbf{h}$, $\mathbf{g}_{\mathbf{x}}$ and $\mathbf{g}_{\mathbf{u}}$ are smooth functions, for some finite $r \in \mathbb{N}_{\geq 0}$. □

The previous definition means that for a differentially flat system, we can express the state and the input vectors as an algebraic function of the flat output vector and its derivatives, up to a finite order.

The implications of differential flatness are favorable for both motion planning and control. Thanks to differential flatness, one can simplify trajectory planning problems both from a theoretical and practical point of view [10–12]. The capacity to

[6] The notation $\mathbf{x}^{(r)}$ represents the rth time derivative of $\mathbf{x}$, i.e., $\mathbf{x}^{(r)} = d^r\mathbf{x}/dt^r$.

obtain the nominal state and input from the output (and its derivatives) allows to plan directly for the flat output, using simple algebraic methods and efficient algorithm. Indeed, the flat output equations of motion are simpler, and in the case of bounds and constraints on the state or input, those can be transformed into constraints on the flat outputs and its derivatives. Although this might produce complex nonlinear constraints on the flat output, one can approximate them with simpler functions with the cost of obtaining a sub-optimal solution, but solving the planning problem in a more efficient way. For example, this method has been successfully employed for the design of a kinodynamic motion planner for an unidirectional-thrust aerial vehicle in a cluttered environment [13].

Furthermore, the knowledge of the nominal state and control input required to follow a certain desired flat output trajectory, can be exploited to design robust controllers [9, 14]. For example, this approach was also used to design a decentralized controller for an aerial manipulator [15].

2.3 Dynamic Feedback Linearizing Control

One very common control method for nonlinear systems to solve tracking control problem is the *feedback linearization* [16–18]. The concept of this method consists on finding a particular output, called *linearizing output* and a control law that linearizes the input-output relation, providing a linear system equivalent to the original one. A standard linear controller can be then applied to the latter equivalent linear system in order to track the desired output trajectory. In the following we shall briefly recall how to practically apply this control method. For more details we refer the reader to more specific books on nonlinear systems as [19–21].

Let us consider the nonlinear system

$$\dot{\mathbf{x}} = \mathbf{f}(\mathbf{x}) + \mathbf{g}(\mathbf{x})\mathbf{u} \tag{2.9a}$$

$$\mathbf{y} = \mathbf{h}(\mathbf{x}), \tag{2.9b}$$

with state vector $\mathbf{x} \in \mathbb{R}^n$, input vector $\mathbf{u} \in \mathbb{R}^m$, output vector $\mathbf{y} = [y_1 \; \dots \; y_m]^\top \in \mathbb{R}^m$, where $\mathbf{f}$, $\mathbf{g}$ and $\mathbf{h}$ are smooth functions. From a practical point of view, in order to feedback linearize the system, one has to differentiate every entry of the output until the input appears, i.e., until we can write

$$\left[y_1^{(r_1)} \; \dots \; y_m^{(r_m)}\right]^\top = \mathbf{b}(\mathbf{x}) + \mathbf{E}(\mathbf{x})\mathbf{u}, \tag{2.10}$$

where $\mathbf{b}(\mathbf{x}) \in \mathbb{R}^m$ collects all the terms that do not depend on the input, and $\mathbf{E} \in \mathbb{R}^{m \times m}$ is called *decoupling matrix*. If the decoupling matrix is invertible over a certain region, the control law

$$\mathbf{u} = \mathbf{E}(\mathbf{x})^{-1}(-\mathbf{b}(\mathbf{x}) + \mathbf{v}), \tag{2.11}$$

where $\mathbf{v} \in \mathbb{R}^m$ is a new *virtual input*, yields to the simpler linear system

$$\begin{bmatrix} y_1^{(r_1)} & \dots & y_m^{(r_m)} \end{bmatrix}^\top = \mathbf{v}. \tag{2.12}$$

r_i is called the *relative degree* of the ith output entry, and we define $r = \sum_{i=1}^m r_i$ as the *total relative degree*. If the total relative degree is equal to the dimension of the system, i.e., $r = n$, then the system is *exactly feedback linearizable*, i.e., (2.12) is equivalent to the original nonlinear system (2.9) and there is no internal dynamics.

Without loss of generality, let us assume that $\mathbf{E}$ is always not invertible because some of its columns are zero.[7] In particular, let the jth column of $\mathbf{E}$ equal to zero. In other words, this means that the input u_j appears in none output entry. In these cases, in order to make u_j appear, one can apply a *dynamic extension* to the other inputs to delay their appearance in the output derivatives. In details, one can consider the new control input $\bar{\mathbf{u}} \in \mathbb{R}^m$ such that $\bar{u}_i = \dot{u}_i$ if $i \neq j$, and $\bar{u}_i = u_j$ for $i = j$. Now the output has to be differentiated one more time to see the input appear:

$$\begin{bmatrix} y_1^{(r_1+1)} & \dots & y_m^{(r_m+1)} \end{bmatrix}^\top = \bar{\mathbf{b}}(\mathbf{x}) + \bar{\mathbf{E}}(\mathbf{x})\bar{\mathbf{u}}. \tag{2.13}$$

If the new decoupling matrix is invertible and the total relative degree is equal to the system dimension plus the new controller states, then the system is said *dynamic feedback linearizable*. If so, $\bar{\mathbf{u}}$ can be designed similarly to (2.11) to obtain an equivalent linear dynamics as in (2.12). The original inputs u_i can be obtained by integration of $\bar{u}_i$, for $i \neq j$. Notice that the presence of the integrals makes the controller "dynamic".

The tracking of any given desired trajectory, $y_i^d(t)$ for $i = 1, \dots, m$ can be achieved applying any linear control technique to the equivalent linear system (2.12). E.g., it is sufficient to use as outer loop a simple controller based on the pole placing technique. Setting the virtual control inputs as

$$v_i = y_i^{d(r_i)} + \sum_{j=0}^{r_i-1} k_{ij}\xi_{ij}, \tag{2.14}$$

where $\xi_{ij} = y_i^{d(j)} - y_i^{(j)}$. One can set the poles of the error dynamics through the gains $k_{ij} \in \mathbb{R}_{>0}$ and for $j = 0, \dots, r_i$ and $i = 1, \dots, m$ to obtain a sufficiently fast exponentially tracking of the desired trajectories. Notice that an explicit measurement of the output and its derivatives is *not* needed at all, since they are algebraic functions of the state and input.

We remark that this method is strongly model based. For this reason, according to the specific system, it might result not robust to model uncertainties. Nevertheless,

[7] If $\mathbf{E}$ is not invertible one can always apply an invertible, state-dependent, input transformation that zeroes the maximum number of columns in $\mathbf{E}$.

the additional linear controller helps in reducing those negative effects. Furthermore, for some complex systems, the control law (2.11) might result very complicate to implement due to the inversion of $\bar{\mathbf{E}}$ and the possible presence of dynamic extensions.

2.4 High Gain Observer

As shown in the previous section, in order to implement the control action, the knowledge of the state of the system is needed. However, measuring the whole state $\mathbf{x}$ using many sensors is often practically unfeasible due to, e.g., the costs and payload limitations, in particular for aerial robots. Furthermore, possible sensor failures call for the ability to still control the platform with a forcedly limited number of sensors.

In order to solve nonlinear observation problems there are mainly two classes of methods: *approximate nonlinear observers* and *exact nonlinear observers*. The first class relies on approximating the nonlinearities with linear or almost-linear maps around the current estimate, the main disadvantage being the local approximative nature of the methods. The second class of methods consists in nonlinear systems whose state is analytically proven to converge to the real state of the original system. Designing such observers is in general much more difficult since it is often hard to prove the asymptotic stability of a nonlinear system. However the observers of this class may guarantee (almost) global convergence. This is why in this thesis we decided to search for an observer in the second class.

In the literature of exact nonlinear observers an important role is played by a particular class of systems known as *in the canonical form*. This is the class of nonlinear systems (2.9) that can be transformed in a triangular form, as:

$$\dot{\mathbf{x}} = \underbrace{\begin{bmatrix} 0 & 1 & \cdots & \cdots & 0 \\ 0 & 0 & 1 & \cdots & 0 \\ \vdots & & & & \vdots \\ 0 & \cdots & \cdots & 0 & 1 \\ 0 & \cdots & \cdots & \cdots & 0 \end{bmatrix}}_{\mathbf{A}} \mathbf{x} + \underbrace{\begin{bmatrix} 0 \\ 0 \\ \vdots \\ 0 \\ 1 \end{bmatrix}}_{\mathbf{B}} \phi(\mathbf{x}, \mathbf{u}) + \boldsymbol{\lambda}(\mathbf{u}) \tag{2.15a}$$

$$w = \underbrace{\begin{bmatrix} 1 & 0 & \cdots & \cdots & 0 \end{bmatrix}}_{\mathbf{C}} \mathbf{x}, \tag{2.15b}$$

where $w \in \mathbb{R}$ is the measurement and $\phi : \mathbb{R}^n \times \mathbb{R}^m \to \mathbb{R}$, $\boldsymbol{\lambda} : \mathbb{R}^m \to \mathbb{R}^n$ are any nonlinear map. For this class of nonlinear systems, in order to estimate the state one can use the nonlinear *High Gain Observer* (HGO) [20]:

$$\dot{\hat{\mathbf{x}}} = \mathbf{A}\hat{\mathbf{x}} + \mathbf{B}\phi(\hat{\mathbf{x}}, \mathbf{u}) + \boldsymbol{\lambda}(\mathbf{u}) + \mathbf{H}(w - \mathbf{C}\hat{\mathbf{x}}), \tag{2.16}$$

with $\mathbf{H} = [\frac{\alpha_1}{\epsilon} \; \frac{\alpha_2}{\epsilon^2} \; \dots \; \frac{\alpha_n}{\epsilon^n}]^\top$ and $\epsilon \in \mathbb{R}_{>0}$. If $\alpha_i \in \mathbb{R}_{>0}$ are set such that the roots of $p^n + \alpha_1 p^{n-1} + \dots + \alpha_{n-1} p + \alpha_n$ have negative real part, then (2.16) ensures almost global convergence of the estimated state to the real one.

Furthermore, let us assume that an output feedback controller $\mathbf{u} = \mathbf{\Gamma}(\mathbf{x}, \mathbf{v})$ (as (2.11) or its dynamic version) is applied to the system. Then one can show that there exists $\epsilon^\star > 0$ such that, for every $0 < \epsilon < \epsilon^\star$, the closed loop system with controller $\mathbf{u} = \mathbf{\Gamma}(\hat{\mathbf{x}}, \mathbf{v})$ and observer (2.16) is exponentially convergent.

However, we recall that, due to the possibly high values of the gains, this observer might suffer from peaking phenomenon during the transient and noise sensitivity. To mitigate those problems, many common practical solutions have been presented in the literature, see e.g., [20, 22]. For example, to overcome the peaking phenomenon, it is sufficient to saturate the estimated state on a bounded region that defines the operative state space bounds for the system in exam. In the presence of measurement noise, the use of a switching-gain approach can guarantee a quick convergence to the real state during the first phase while reducing the noise effects at steady state.

References

1. Isidori, A., Moog, C.H., De Luca, A.: A sufficient condition for full linearization via dynamic state feedback. In: 25th IEEE Conference on Decision and Control, vol. 25, pp. 203–208. IEEE (1986)
2. Siciliano, B., Sciavicco, L., Villani, L., Oriolo, G.: Robotics: Modelling, Planning and Control. Springer (2009)
3. Siciliano, B., Khatib, O.: Handbook of Robotics. Springer (2008)
4. Spong, M.W., Hutchinson, S., Vidyasagar, M.: Robot Modeling and Control, vol. 3. Wiley, New York (2006)
5. Lynch, K.M., Park, F.C.: Modern Robotics: Mechanics, Planning, and Control. Cambridge University Press (2017)
6. Corke, P.: Robotics, Vision and Control: Fundamental Algorithms in MATLAB® Second, Completely Revised, vol. 118. Springer (2017)
7. Fliess, M., Lévine, J., Martin, P., Rouchon, P.: Flatness and defect of nonlinear systems: introductory theory and examples. Int. J. Control **61**(6), 1327–1361 (1995)
8. Murray, R.M., Rathinam, M., Sluis, W.: Differential flatness of mechanical control systems: a catalog of prototype systems. In: ASME International Mechanical Engineering Congress and Exposition, San Francisco, CA, Nov (1995)
9. Rigatos, G.G.: Nonlinear Control and Filtering Using Differential Flatness Approaches: Applications to Electromechanical Systems, vol. 25. Springer (2015)
10. Rouchon, P., Fliess, M., Levine, J., Martin, P.: Flatness and motion planning: the car with n trailers. In: 1993 European Control Conference, pp. 1518–1522, Groningen (1993)
11. Chamseddine, A., Zhang, Y., Rabbath, C.A., Join, C., Theilliol, D.: Flatness-based trajectory planning/replanning for a quadrotor unmanned aerial vehicle. IEEE Trans. Aerosp. Electron. Syst. **48**(4), 2832–2848 (2012)
12. De Luca, A., Oriolo, G.: Trajectory planning and control for planar robots with passive last joint. Int. J. Robot. Res. **21**(5–6), 575–590 (2002)
13. Boeuf, A., Cortés, J., Alami, R., Siméon, T.: Enhancing sampling-based kinodynamic motion planning for quadrotors. In: 2015 IEEE/RSJ International Conference on Intelligent Robots and Systems, pp. 2447–2452, Hamburg, Germany, Sept (2015)

14. Tang, C.P., Miller, P.T., Krovi, V.N., Ryu, J., Agrawal, S.K.: Differential-flatness-based planning and control of a wheeled mobile manipulator—theory and experiment. IEEE/ASME Trans. Mechatron. **16**(4), 768–773 (2011)
15. Tognon, M., Yüksel, B., Buondonno, G., Franchi, A.: Dynamic decentralized control for protocentric aerial manipulators. In: 2017 IEEE International Conference on Robotics and Automation, pp. 6375–6380, Singapore, May (2017)
16. Oriolo, G., De Luca, A., Vendittelli, M.: WMR control via dynamic feedback linearization: design, implementation, and experimental validation. IEEE Trans. Control Syst. Technol. **10**(6), 835–852 (2002)
17. De Luca, A.: Decoupling and feedback linearization of robots with mixed rigid/elastic joints. Int. J. Robust Nonlinear Control: IFAC-Affil. J. **8**(11), 965–977 (1998)
18. Martin, P., Devasia, S., Paden, B.: A different look at output tracking: control of a VTOL aircraft. In: Proceedings of the 33rd IEEE Conference on Decision and Control, vol. 3, pp. 2376–2381. IEEE (1994)
19. Slotine, J.J.E., Li, W.: Applied Nonlinear Control. Prentice Hall (1991)
20. Khalil, H.K.: Nonlinear Systems, 3rd edn. Prentice Hall (2001)
21. Isidori, A.: Nonlinear Control Systems, 3rd edn. Springer (1995)
22. Ahrens, J.H., Khalil, H.K.: High-gain observers in the presence of measurement noise: a switched-gain approach. Automatica **45**(4), 936–943 (2009)

Chapter 3
Model of the Robotic Elements

3.1 Tethering Link

In this book we address the control problem of tethered aerial vehicles from the most generic point of view, deriving general and fundamental theoretical results that can be then easily applied to several practical cases. For this reason, in Sect. 4.3 we consider the aerial vehicle tethered to a moving platform by a *generic link*, either a cable, a chain, a rope, a bar or a strut. All the considered links can be divided into three major categories, schematically represented in Fig. 3.1:

(a) Links that can be *only stretched*, i.e., that can support *only tensions*. To this category fall all the types of cable-like link, e.g., a chain, a rope, etc.
(b) Links that can be *only compressed*, i.e., that can support *only compressions*. To this category fall all the types of strut-like link, such as a pneumatic suspension.
(c) Links that can be *both stretched and compressed*, i.e., that can support *both tensions and compressions*. To this category fall all the types of bar-like link, such as a beam, a pole, etc.

The main variables that describe a generic link are: the position of the edges given by the vectors $\mathbf{p}_1 \in \mathbb{R}^3$ and $\mathbf{p}_2 \in \mathbb{R}^3$, the unstressed length $l_0 \in \mathbb{R}_{>0}$ and the intensity of the internal force $f_L \in \mathbb{R}$. Other equivalent but still meaningful variables can be defined: the length of the link $l = \|\mathbf{p}_1 - \mathbf{p}_2\| \in \mathbb{R}$ and the normalized axis of the link $\mathbf{d} = (\mathbf{p}_1 - \mathbf{p}_2)/l \in \mathbb{R}^3$. Figure 3.2 shows the main variables describing a link.

When the link is pulled the internal force is called *tension* and $f_L > 0$, whereas when it is compressed the internal force is called *compression* and $f_L < 0$. When $f_L = 0$ the link is *slack*. The easiest way to model the link is as a hybrid system with two states: slack or non-slack, i.e., taut/compressed. When the link is slack and $f_L = 0$, the length of the link can be: (a) $l \leq l_0$ for a cable-like link; (b) $l \geq l_0$ for a strut-like link; and (c) $l = l_0$ for a bar-like link. In this condition the two ends of the cable are treated as two independent systems, as done in [1, 2]. However, we are not interested in this case since we shall design a controller that will keep away the system from the slack state for the cable and strut-like link cases.

M. Tognon and A. Franchi, *Theory and Applications for Control of Aerial Robots in Physical Interaction Through Tethers*, Springer Tracts in Advanced Robotics 140,
https://doi.org/10.1007/978-3-030-48659-4_3

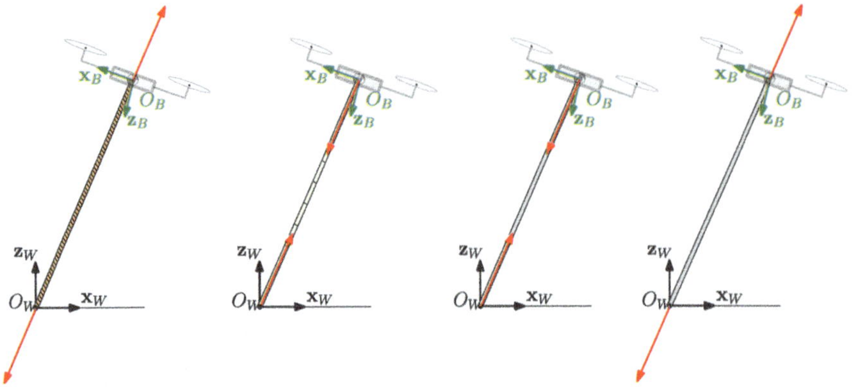

(a) Cable-like link can be only stretched.

(b) Strut-like link can be only compressed

(c) Bar-like link can be compressed (left) and stretched (right)

Fig. 3.1 Three type of considered link. The red arrows indicate the external forces (or reaction forces) that stretch or compress the link, according to the category

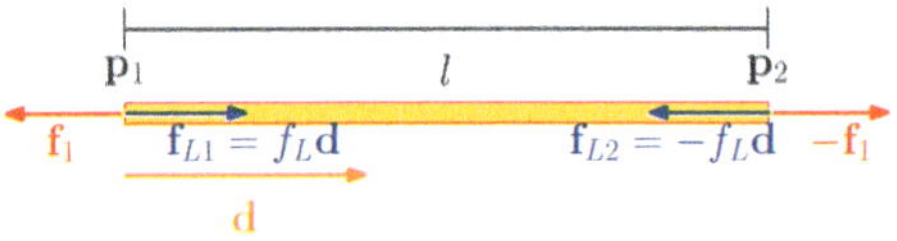

Fig. 3.2 Schematic representation of a general link and its main variables

On the other hand, when the cable is taut or compressed, i.e., $f_L \neq 0$, we assume that the length of the link remain constant independently from the internal force, $l = l_0$. This is equivalent to assume that, in the domain of interest, a non-slack link is a rigid element whose elasticity and deformations are negligible. This assumption is very common in the related literature review [3, 4], and is valid as soon as the maximum internal force of interest is much smaller than the stiffness coefficient of the link. Under this assumption we have that the internal force at the two ends are opposite, and always along the link axis, $\mathbf{f}_{L2} = -\mathbf{f}_{L1} = -f_L\mathbf{d}$.

As also commonly done in the related state of the art, we assume negligible mass and inertia of the link with respect to the one of the aerial vehicle. Also this assumption is easily met using very lightweight links like kite cables or link structures based on carbon fiber.

Nevertheless, one could use more complex models to describes all the previously neglected effects. For example a *spring-damper* system can be used to better describe the deformations of the link subjected to external forces. According to the particular type of link, even more complex and accurate models could be employed. For example, in the case of a cable-like link, one can use a *Standard Linear Solid model* (SLS) [5–7] consisting of a series of a spring and a parallel of spring-damper. This model better describes the response delay due to the relatively slow microscopic deformation process acting in the rope when some external forces are stretching it. Furthermore, in order to also model flexibility of cables and the bending due to gravity, one could model it by a *finite element approximation* [7]. It consists on modeling the link as a chain of elements, i.e., as a finite number of smaller links connected in series, as done in [8, 9].

Although the previously mentioned models better describe the real behaviors of the considered links, actually, the effects that they additionally describe with respect to the simple mass-less rigid body model, are negligible in the domain of interest. Furthermore, they are specific only for certain type of links. On the other hand, the mass-less rigid body model is a general model that can be applied to the several classes of link previously presented. Thus it is more suitable for the aimed control objectives. This is why, for the sake of designing control and observer methods, we choose this model to derive the equations of motion of the considered tethered aerial vehicle (see Sect. 4.3). Nevertheless, in Sect. 5.4.2 we shall show that the proposed methods based on the mass-less rigid body model are robust enough also considering more accurate non-ideal models.

3.2 Unidirectional Thrust Vehicles

As usual, we start by defining an inertial world frame $\mathcal{F}_W = \{O_W, \mathbf{x}_W, \mathbf{y}_W, \mathbf{z}_W\}$ where O_W is its origin, placed arbitrarily, and $(\mathbf{x}_W, \mathbf{y}_W, \mathbf{z}_W)$ are the orthogonal unit vectors. We consider $\mathbf{z}_W$ parallel and opposite to the gravity vector. Then we define the body frame $\mathcal{F}_R = \{O_R, \mathbf{x}_R, \mathbf{y}_R, \mathbf{z}_R\}$ rigidly attached to the vehicle and centered in O_R, the vehicle CoM. We consider $\mathbf{z}_R$ parallel and opposite to the total thrust vector. The position of O_R and orientation of $\mathcal{F}_R$ w.r.t. $\mathcal{F}_W$ are described by the vector $\mathbf{p}_R \in \mathbb{R}^3$ and the rotation matrix $\mathbf{R}_R \in \mathsf{SO}(3)$, respectively. Then we denote by the vector $\boldsymbol{\omega}_R \in \mathbb{R}^3$ the angular velocity of $\mathcal{F}_R$ w.r.t. $\mathcal{F}_W$ and expressed in $\mathcal{F}_R$. The variables describing the vehicle are depicted in Fig. 3.3.

As already announced in Sect. 2.1.3, the vehicle is modeled as a rigid body with mass $m_R \in \mathbb{R}_{>0}$ and moment of inertia about O_R, defined w.r.t. $\mathcal{F}_R$, described by the

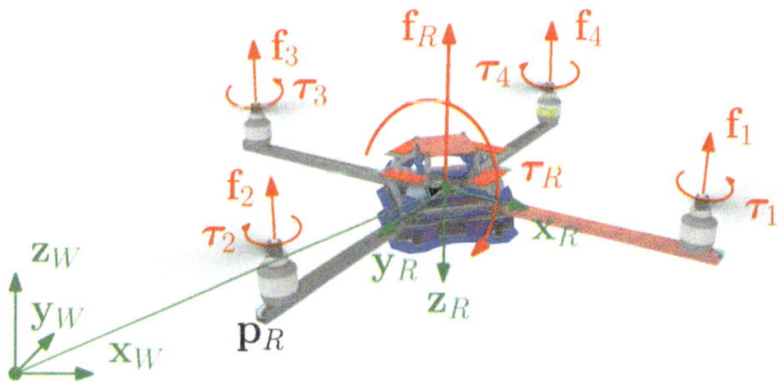

Fig. 3.3 Schematic representation of a collinear multirotor and its main quantitites. Although the vehicle is represented as a quadrotor, actually it can be any collinear multirotor, such as an hexarotor, octorotor, etc.

positive definite matrix $\mathbf{J}_R \in \mathbb{R}^{3\times3}_{>0}$ [10, 11]. The motion of the vehicle is controlled by the coordinated action of four control inputs: *(i)* $f_R \in \mathbb{R}_{\geq 0}$ is the *intensity of the total thrust* applied in O_R such that $\mathbf{f}_R = -f_R\mathbf{z}_R$, which generates translational motion, and *(i)* $\boldsymbol{\tau}_R = [\tau_{Rx}\ \tau_{Ry}\ \tau_{Rz}]^\top \in \mathbb{R}^3$ is the *total moment* applied to $\mathcal{F}_R$ and expressed in $\mathcal{F}_R$, which generates rotational motion.

Similarly to Sect. 2.1.3, the dynamics of the system is computed applying the Newton-Euler equations, thus obtaining $\dot{\mathbf{R}}_R = \mathbf{R}_R\boldsymbol{\Omega}_R$, and

$$m_R\ddot{\mathbf{p}}_R = -m_R g\mathbf{e}_3 - f_R\mathbf{R}_R\mathbf{e}_3 \tag{3.1a}$$

$$\mathbf{J}_R\dot{\boldsymbol{\omega}}_R = -\boldsymbol{\omega}_R \times \mathbf{J}_R\boldsymbol{\omega}_R + \boldsymbol{\tau}_R. \tag{3.1b}$$

This model is general and well describes the dynamics of the majority of unidirectional thrust aerial vehicles as ducted fan and multirotors vehicles with four or more rotors. Indeed, (3.1) encapsulates all the nonlinearities and the underactuation of unidirectional-thrust aerial vehicles. One can notice how, in order to apply a certain acceleration, the vehicle has to be oriented such that the total thrust vector is oriented such to ensure an acceleration that is equal to the given acceleration plus the gravity compensation. This shows the coupling between translational and rotational dynamics.

Thanks to its generality, we will use this model to describe the dynamics of the aerial vehicles considered in the following. However, how to practically generate the total thrust and the torque will be treated in the following section.

3.3 Actuators

3.3.1 *Thrusters*

In this thesis we consider a particular class of unidirectional-thrust aerial vehicles. We consider aerial vehicles that generate the total thrust and torque by the aerodynamic forces and moments in turn generated by multiple collinear propellers. In practice, a model for those systems is needed to map total thrust and torque into the real control inputs.

Let us assume that the vehicle is endowed with $n \in \mathbb{N}_{\geq 4}$ thrusters. The generic i-th thruster is rigidly attached to the main frame oriented as $-\mathbf{z}_R$, and its position is given by the vector $\mathbf{b}_i \in \mathbb{R}^3$ with respect to $\mathcal{F}_R$. It is composed by a pair of brushless-motor plus propeller. Making the propeller spin at a certain velocity $w_i \in \mathbb{R}_{\geq 0}$, it can produce a force $f_i\mathbf{z}_R$, whose intensity is equal to $f_i = c_f w_i^2$, where $c_f \in \mathbb{R}_{>0}$ is called *lift factor* and depends on the aerodynamic properties of the propeller blades [10, 12–15]. When a propeller is spinning, the resistance of the air generates some horizontal forces on the blade, as well. Those drag forces, multiplied by the momentum arm and integrated over the rotor, generate a moment about the rotor shaft, that in the aerial robotics community is normally called *drag moment*. A reaction torque acts on the rotor in the opposite direction of rotation of the propeller. The latter in turn generates a torque on the main frame of the vehicle that results to be proportional to the square of the propeller angular velocity, i.e., $\boldsymbol{\tau}_i = c_i c_\tau w_i^2 \mathbf{e}_3$, where *(i)* $c_i = 1$ ($c_i = -1$) if the i-th propeller angular velocity vector has the same direction of $\mathbf{z}_R$ ($-\mathbf{z}_R$), i.e., the propeller spins CCW (CW) when watched from its top; *(ii)* $c_\tau \in \mathbb{R}_{>0}$ also depends on the aerodynamic properties of the propeller.

Finally the total thrust and torque applied to the vehicle frame are given by:

$$f_R = \sum_{i=1}^{n} c_f w_i^2 \tag{3.2a}$$

$$\boldsymbol{\tau}_R = \sum_{i=1}^{n} (c_f \mathbf{b}_i \times \mathbf{e}_3 + c_i c_\tau \mathbf{e}_3) w_i^2. \tag{3.2b}$$

In particular, for a quadrotor-like vehicle where, $n = 4$, $\mathbf{b}_i = b[c_{\alpha_i}\ s_{\alpha_i}\ 0]^\top$ with $b \in \mathbb{R}_{>0}$ and $\alpha = (i-1)\pi/2$, $c_i = (-1)^i$ and $i = 1, \ldots, 4$, we have that

$$\begin{bmatrix} f_R \\ \boldsymbol{\tau}_R \end{bmatrix} = \underbrace{\begin{bmatrix} c_f & c_f & c_f & c_f \\ 0 & -c_f b & 0 & c_f b \\ c_f b & 0 & -c_f b & 0 \\ -c_\tau & c_\tau & -c_\tau & c_\tau \end{bmatrix}}_{\mathbf{F}} \begin{bmatrix} w_1^2 \\ w_2^2 \\ w_3^2 \\ w_4^2 \end{bmatrix}. \tag{3.3}$$

Notice that the allocation matrix $\mathbf{F}$ is square and full rank, thus always invertible. Once the desired total thrust and torque are computed, we can compute the spinning velocity of each propeller that should be actuated to generate the desired control action just by inverting (3.3).

In particular conditions such as strong wind or at a very high speed, other gyroscopic and aerodynamic effects such as drag and blade flapping should be considered [10, 12, 16]. Additional aerodynamic effects are the ground and ceiling effects that arise whenever the vehicle flies close to a surface. However, as normally done in the related literature, we do not consider those effects since they are negligible in the domain of interest.

Furthermore, notice that the actuation model presented in (3.2) assumes that the motors can actuate the desired spinning velocity instantaneously. Nevertheless, changing the spinning velocity instantaneously would require an infinite torque that is obviously practically unfeasible. One should instead add to (3.2) the dynamics of the motor, both from a mechanic and electronic point of view [17], together with the dynamics of the electronic speed controller (ESC). A system identification of the overall closed loop system can be done to estimate the model parameters. Nevertheless, for control design purposes, we can assume that the spinning velocity variations are limited in the domain of interest. Under this assumption and thanks to the employed brushless controller [18] that guarantees minimal response times, model (3.2) results a good approximation of the real behavior. On the other hand, more complex and realistic models, such as the ones mentioned before, can be used for the validation of the theoretical results in a more realistic condition (see Chap. 5).

3.3.2 *Link Actuator*

In this thesis the link actuator is modeled as a cylinder that transform the rotational motion of a motor to a translational motion, namely the variation of the link length. In particular, in the case of a cable-like link, the cable is rolled on the cylinder, while in the case of a bar-like link, the cylinder is a gear mechanism that moves back and forward the link changing its length. The cylinder is moved by a motor that exerts an input torque $\tau_W \in \mathbb{R}$ about the longitudinal axis of the cylinder. We assume that the rotational inertia and radius of the link actuator, denoted by $J_W \in \mathbb{R}_{>0}$ and $r_W \in \mathbb{R}_{>0}$, respectively, are constant in the domain of interest. The dynamic of the link actuator is

$$J_W \ddot{\vartheta}_W = \tau_W + f_L r_W, \tag{3.4}$$

where $\ddot{\vartheta}_W$ is the angular acceleration of the actuator. Since we are more interested in the dynamics of the link length, we can easily describe it from (3.4). Assuming no backlash we can write $l = r_W \vartheta_W$, thus

$$\bar{J}_W \ddot{l} = \bar{\tau}_W + f_L, \tag{3.5}$$

where $\bar{J}_W = J_W / r_W^2$ and $\bar{\tau}_W = \tau_W / r_W$.

According to the real implementation some assumptions, like the constant inertia and radius might not hold anymore. However the model can be easily changed accordingly.

3.4 Sensory Setup

To control the previous modeled aerial vehicles the knowledge of the state is needed in most of the cases. In other words, the position, the linear velocity, the attitude and the angular velocity of the vehicle have to be estimated from the available sensors, in order to then compute the control action.

The most basic sensor available on practically all vehicles is the *inertial measurement unit* (IMU) [10, 19, 20]. It normally consists of an 3-axes accelerometer and a 3-axes gyroscope. The first measures the so call "specific acceleration", namely the acceleration of the vehicle with respect to the body frame minus the gravity vector. Defining $\mathbf{w}_{acc} \in \mathbb{R}^3$ the measure coming from the accelerometer, and assuming that the IMU is calibrated, centered at O_R and its axes are aligned with the ones of $\mathcal{F}_R$, we have that:

$$\mathbf{w}_{acc} = \mathbf{R}_R^\top(\ddot{\mathbf{p}}_R + g\mathbf{e}_3). \tag{3.6}$$

On the other hand, the gyroscope, whose measurements are defined by the vector $\mathbf{w}_{gyr} \in \mathbb{R}^3$, directly measures, under the same assumptions, the angular velocity of $\mathcal{F}_R$ with respect to $\mathcal{F}_W$ expressed in $\mathcal{F}_R$:

$$\mathbf{w}_{gyr} = \boldsymbol{\omega}_R. \tag{3.7}$$

These two quantities are not enough to estimate the full time-varying attitude, because the rotation along $\mathbf{z}_R$, is not observable in hovering conditions [10, 21]. This is why the IMU is often equipped with a *magnetomer* that measures the ambient magnetic field with respect to the body frame. In the absence of disturbances, the latter corresponds to the known Earth's magnetic field defined by the vector $\mathbf{h}^W \in \mathbb{R}^3$. Under this assumption, the magnetometer measurement, $\mathbf{w}_{mag} \in \mathbb{R}^3$, is equal to:

$$\mathbf{w}_{mag} = \mathbf{R}_R^\top \mathbf{h}^W. \tag{3.8}$$

The combination of the previous three sensors is in principle sufficient to estimate the rotational part of the state [22–25], i.e., the attitude and the angular velocity.

For estimating the rest of the state, i.e., the position and the translational velocity, some other exteroceptive sensors are usually needed. Some examples are motion capture system (MoCap) for precise indoor localization, GPS and differential GPS for outdoor environments, and various type of cameras for outdoor GPS-denied environments. Since the design of new localization methods using those kind of exteroceptive sensors is not the focus of this thesis, we will model them as a direct measurement of the configuration of the robot. Thus, defining $\mathbf{w}_{ext}$ the output measurement, we have that:

$$\mathbf{w}_{ext} = (\mathbf{p}_R, \mathbf{R}_R). \tag{3.9}$$

The authors of [10] present the most popular methods to fuse all the mentioned sensors in order to finally obtain a precise estimation of the full state.

However, in Chap. 4 (see Sects. 4.7 and 4.8) we shall show that for a tethered aerial vehicle, those exteroceptive sensors are not needed to estimate the full state of the system. Indeed, we found that, thanks to the link constraint, in a 2D environment only an IMU is enough to retrieve an estimation of the fulls state (see Sect. 4.8). On the other hand, in the 3D environment, we found that the minimal sensory setup consists of a standard IMU and magnetometer, plus only some encoders to measure the attitude of the link and its length (if not constant).

In order to model the measurement of an encoder let us define two frames $\mathcal{F}_1 = \{O_1, \mathbf{x}_1, \mathbf{y}_1, \mathbf{z}_1\}$ and $\mathcal{F}_2 = \{O_2, \mathbf{x}_2, \mathbf{y}_2, \mathbf{z}_2\}$ such that $O_1 = O_2$ and $\mathbf{x}_1 = \mathbf{x}_2$. Then we have that:

$$\mathbf{w}_{enc} = \theta, \tag{3.10}$$

where $\theta \in \mathbb{R}$ is the angle to make $\mathcal{F}_2$ coincide with $\mathcal{F}_1$ rotating it about $\mathbf{y}_1$.

The models presented so far are the ideal ones and they will be used in the following to design deterministic and almost globally convergent nonlinear observers. However, in practice every sensor is affected by noise and biases. We will then rely on the proven robustness of the designed observer to deal with those non-idealities (see Chap. 5). Another approach would be to design a stochastic estimators that deals with noisy measurements. However they are normally based on linear approximation of the models and therefore they are not globally convergent.

References

1. Sreenath, K., Kumar, V.: Dynamics, control and planning for cooperative manipulation of payloads suspended by cables from multiple quadrotor robots. In: Robotics: Science and Systems, Berlin, Germany, June (2013)
2. Sreenath, K., Michael, N., Kumar, V.: Trajectory generation and control of a quadrotor with a cable-suspended load-a differentially-flat hybrid system. In: 2013 IEEE International Conference on Robotics and Automation (ICRA), pp. 4888–4895. IEEE (2013)
3. Lupashin, S., D'Andrea, R.: Stabilization of a flying vehicle on a taut tether using inertial sensing. In: 2013 IEEE/RSJ International Conference on Intelligent Robots and Systems, pp. 2432–2438, Tokyo, Japan, Nov (2013)
4. Nicotra, M.M., Naldi, R., Garone, E.: Nonlinear control of a tethered UAV: the taut cable case. Automatica **78**, 174–184 (2017)
5. Sandino, L.A., Bejar, M., Kondak, K., Ollero, A.: Advances in modeling and control of tethered unmanned helicopters to enhance hovering performance. J. Intell. Robot. Syst. **73**(1–4), 3–18 (2014)
6. Liu, H.P., Anderson, D.L., Kanamori, H.: Velocity dispersion due to anelasticity; implications for seismology and mantle composition. Geophys. J. Int. **47**(1), 41–58 (1976)
7. Zienkiewicz, O.C., Taylor, R.L.: The Finite Element Method for Solid and Structural Mechanics. Elsevier (2005)
8. Lee, T.: Geometric controls for a tethered quadrotor UAV. In: 2015 54th IEEE Conference on Decision and Control (CDC), pp. 2749–2754, Osaka, Japan, Dec (2015)
9. Goodarzi, F.A., Lee, T.: Stabilization of a rigid body payload with multiple cooperative quadrotors. J. Dyn. Syst. Meas. Control **138**(12), 121001–121001–17 (2016)
10. Mahony, R., Kumar, V., Corke, P.: Multirotor aerial vehicles: modeling, estimation, and control of quadrotor. IEEE Robot. Autom. Mag. **19**(3), 20–32 (2012)
11. Hua, M.-D., Hamel, T., Morin, P., Samson, C.: A control approach for thrust-propelled underactuated vehicles and its application to VTOL drones. IEEE Trans. Autom. Control **54**(8), 1837–1853 (2009)
12. Hamel, T., Mahony, R., Lozano, R., Ostrowski, J.: Dynamic modelling and configuration stabilization for an X4-Flyer. In: 15th IFAC World Congress, pp. 217–222, Barcelona, Spain (2002)
13. Pounds, P.E., Mahony, R., Corke, P.: Modeling and control of a large quadrotor robot. Control Eng. Pract. **18**(7), 691–699 (2010)
14. Valavanis, K.P.: Advances in Unmanned Aerial Vehicles: State of the Art and the Road to Autonomy. Intelligent Systems, Control and Automation: Science and Engineering, vol. 33. Springer (2007)
15. Pucci, D., Hamel, T., Morin, P., Samson, C.: Nonlinear control of aerial vehicles subjected to aerodynamic forces. In: 2013 IEEE 52nd Annual Conference on Decision and Control (CDC), pp. 4839–4846. IEEE (2013)

16. Faessler, M., Franchi, A., Scaramuzza, D.: Differential flatness of quadrotor dynamics subject to rotor drag for accurate tracking of high-speed trajectories. IEEE Robot. Autom. Lett. **3**(2), 620–626 (2018)
17. Bangura, M., Mahony, R.: Thrust control for multirotor aerial vehicles. IEEE Trans. Robot. **33**(2), 390–405 (2017)
18. Franchi, A., Mallet, A.: Adaptive closed-loop speed control of BLDC motors with applications to multi-rotor aerial vehicles. In: 2017 IEEE International Conference on Robotics and Automation, pp. 5203–5208, Singapore, May (2017)
19. Ahmad, N., Ghazilla, R.A.R., Khairi, N.M., Kasi, V.: Reviews on various inertial measurement unit (IMU) sensor applications. Int. J. Signal Process. Syst. **1**(2), 256–262 (2013)
20. Martin, P., Salaün, E.: The true role of accelerometer feedback in quadrotor control. In: 2010 IEEE International Conference on Robotics and Automation, pp. 1623–1629, Anchorage, AK, May (2010)
21. Leishman, R.C., Macdonald, J.C., Beard, R.W., McLain, T.W.: Quadrotors and accelerometers: state estimation with an improved dynamic model. IEEE Control Syst. **34**(1), 28–41 (2014)
22. Mahony, R., Hamel, T., Pflimlin, J.-M.: Nonlinear complementary filters on the special orthogonal group. IEEE Trans. Autom. Control **53**(5), 1203–1218 (2008)
23. Kraft, E.: A quaternion-based unscented kalman filter for orientation tracking. In: Proceedings of the Sixth International Conference of Information Fusion, vol. 1, pp. 47–54 (2003)
24. Scandaroli, G.G., Morin, P., Silveira, G.: A nonlinear observer approach for concurrent estimation of pose, IMU bias and camera-to-IMU rotation. In: 2011 IEEE/RSJ International Conference on Intelligent Robots and Systems (IROS), pp. 3335–3341 (2011)
25. Scandaroli, G.G., Morin, P.: Nonlinear filter design for pose and IMU bias estimation. In: 2011 IEEE International Conference on Robotics and Automation (ICRA), pp. 4524–4530 (2011)

Chapter 4
Theory of Tethered Aerial Vehicles

4.1 Related Works and Problem Statement

Driven by the relevance of this topic, several control and estimation schemes have been presented in the robotic literature. For the case in which the aerial vehicle is an helicopter, the authors of [1] presented a method to land it on the deck of a ship in rough sea using a cable. The controller is based on a time-scale separation technique between the rotational and translational dynamics. In the context of [2], the authors of [3] presented a control scheme based on a PID scheme together with partial model inversion, to stabilize the flight to a constant vertical elevation with the goal of both improving hovering in windy conditions and land on a mobile platform. Notice that a force sensor (load cell) is used to measure the tension along the cable and to compensate its effect on the vehicle. The latter work has been then extended to a tether aerial vehicle together with a winch, in order to perform the landing of a small unmanned helicopter, without the use of GPS sensors but relying on simpler magnetic encoders [4]. Furthermore, in [5] the authors proposed a square-root unscented Kalman filter for the estimation of the attitude and the relative position of the vehicle with respect to the ground anchoring point.

For the case of cable-tethered underactuated multi-rotor system moving on a restricted 2D vertical plane, [6] presents a controller that, under a quasi-static assumption, stabilizes the elevation (angle) of the cable to a constant value using only inertial onboard sensors. In particular, the controller is based on the separation of the translational and rotational dynamics, and on a partial inversion of the elevation dynamics where they fix the total thrust to a value sufficiently high to preserve the tautness of the cable. The state estimator is instead based on a UKF. The authors of [7, 8] also present a controller to stabilize the elevation of a tethered multi-rotor to a constant value while also ensuring the positivity of the cable tension. The controller is based on the combination of a hierarchical cascade approach to stabilize the system and a reference governor to ensure a positive internal tension.

M. Tognon and A. Franchi, *Theory and Applications for Control of Aerial Robots in Physical Interaction Through Tethers*, Springer Tracts in Advanced Robotics 140,
https://doi.org/10.1007/978-3-030-48659-4_4

The presented approaches provide a good basis for observation and control of such systems. However, to improve those methods, overcoming some of their drawbacks, and to extends the results to a more general system, some problems have to to be addressed:

(i) study a more *generic system* in order to enlarge the validity of the results. In particular one could think about three extensions of the classical tethered aerial vehicle: *(i)* Develop a method that can cope with *any kind of link* (i.e., a cable, a strut or a bar), differently from the state-of-the-art methods, which have been designed only for cable links. Indeed the use of rigid link can be beneficial for, e.g., sustain part of the platform weight, or allow a full-bilateral physical interaction with a ground object. *(ii)* Furthermore, instead of considering a static anchoring point, one could consider a *generic platform* moving in the 3D environment, without particular constraints. This allows to consider a very wide class of possible vehicles. *(iii)* Finally one could consider a *link actuator* in order to be able to also change the length of the link, thus controlling the full 3D position of the aerial vehicle with respect to the moving platform.

(ii) *track a time varying trajectory*. While the methods in the literature are based on a quasi-static assumption aiming to a stabilization problem, in many practical cases it is important to track a time-varying trajectory to, e.g., let the aerial robot track a moving target.

(iii) *track a desired link force profile*. In many cases it is important to precisely regulate the force link to any desired value, possibly including negative link force (pushing) if the link is a bar. For example, if the link is a cable, one may want to keep it always taut while at the same time avoiding peaks on the tension that can damage the cable or its attaching mechanisms. If the link is a bar, then one may want to push against it with an appropriate force to perform physical interaction with the environment. However, the methods in the literature are able, in the best case [7], to keep the force link positive, but not at a prescribed value, possibly time-varying.

(iv) *observe the state* from a minimal set of sensors in any dynamic condition. Instead, in the state-of-the-art this is done at best in a quasi-static assumption or with a linearized method [4, 6].

Table 4.1 shows the features of the methods proposed in the state of the art and of the ones proposed in this book.

In order to achieve the previous objectives, a thoroughly analysis of the system from a theoretical point of view is needed. This is why, one of the goal of this book is to provide a fundamental study of the tethered aerial vehicle system, in order to have a clear understanding of its dynamics and the corresponding properties related to control, planning and estimation. Our formal analysis is focused on three main objectives along the axes of control, planning and localization:

(1) Find the output for which it is possible to compute analytically and offline the state of the system while exactly tracking the desired output trajectory, and the nominal inputs required to do so. This option is very useful for control, to compute

Table 4.1 Features of the methods proposed by the state of the art and of the ones proposed in this book

	[3–5]	[6]	[7, 8]	This book
Stabilization of the position	✓	✓	✓	✓
Taut cable ensured			✓	✓
Control of tension and compression				✓
Independent tracking of position and internal force				✓
Independent tracking of position and partial attitude				✓
2D state estimation with IMU only		✓		✓
3D state estimation with IMU plus encoders	✓			✓
Global state estimation				✓

the feed-forward terms, and in a possible pre-planning strategy. To achieve this objective we investigate which are the *flat outputs* and we explicit the algebraic relations that demonstrate the *differential flatness* of the system (see Sect. 4.4).

(2) For each of the found set of flat outputs, given any desired sufficiently smooth trajectory, we aim at designing a control strategy to obtain the exact trajectory tracking (see Sect. 4.6). The control method will be based on the static or dynamic (if needed) *feedback linearization* method presented in Sect. 2.3.

(3) Provide the mathematical tools to implement the control strategy in a real world using a minimal set of typically available on-board sensors. For this goal, we propose a minimal standard sensory equipment and we show how to design an almost global nonlinear observer in order to estimate the state from the available measurements only. This enables the possibility to close the control loop with standard and minimal measurements (see Sects. 4.7 and 4.8).

Based on these results, practical problems coming from applicative scenarios can be solved more easily. An example can be found in Chap. 6 where we exploited the theoretical results to solve the practical but challenging problem of landing and takeoff on/from highly sloped surfaces.

4.2 Contribution

This part of the book provides a thorough analysis of a generic tethered system in which we formally study the dynamics of the system and its intrinsic properties, along the three major axes of control, planning and estimation.

Firstly we consider a much more generic tethered system that extends the previously considered system by four major points. Indeed, we analyze a system firstly composed by an *unidirectional-thrust aerial vehicle flying in the 3D space* (not only in a 2D plane as some of the related publications). The vehicle is attached by a *generic link* to an *independently moving platform*. Differently from the state of the art where only the cable case is considered, here the generic link can be either a cable, a strut or a bar. In this way we can consider links that can be:

(i) only stretched, like cables, that allow only positive internal forces called tensions;
(ii) only compressed, like strut, that allow only negative internal forces called compression;
(iii) both compressed and stretched, like bars, that allow both positive and negative internal forces.

At the other end, the link is attached to a platform that can independently move in the 3D space. Differently from the related literature where the link is anchored to a fixed point, in this book we consider a much generic case in which the link could be attached to a car, a sheep, another aircraft, or even to a human operator.

Finally, for the sake of generality, we consider that a *link actuator* is capable of changing the length of the link applying a certain controllable force. This device can be a winch in the case of a cable-link, or an actuated telescopic mechanism in the case of a bar-link. The modeling of this generic system is provided in Sect. 4.3.

We started our analysis with the investigation of the *differential flatness* property, looking for the sets of flat outputs (if more then one). We proved that the system is differentially flat with respect to two outputs $\mathbf{y}^a$ and $\mathbf{y}^b$.

(a) $\mathbf{y}^a$ is a *force-related output*. It contains the position of the vehicle with respect to the moving platform (expressed in spherical coordinates), the internal force along the link, and the rotation of the vehicle along the thrust axis (see Sect. 4.6.1). This shows that the position and the internal force along the link can be controlled independently. Notice the parallel with hybrid motion/force control of ground manipulators in the presence of kinematic constraints [9].
(b) $\mathbf{y}^b$ is an *attitude-related output*. It contains the position of the vehicle with respect to the moving platform, a particular angle related to the attitude of the vehicle with respect to the one of the link, and the rotation of the vehicle along the thrust axis (see Sect. 4.6.2). This shows that the position and this particularly defined angle can be controlled independently.

These are in turn *flat outputs* for the system. This property fulfills the planning related objective, and allows to know in advance the nominal state and inputs while exactly tracking a desired output trajectory.

For the two sets of flat outputs we then designed a hierarchical controller based on the separation between the rotational and translational dynamics controllers (see Sect. 4.5). We designed this controller with the aim of stabilizing the flat outputs to a desired value, keeping it simple to implement also for commercially available aerial vehicle for which one can control only the attitude or, at best, the angular velocity. However, after the experimental validation, we noticed that this controller cannot provide good tracking performance when the desired trajectory requires high velocities and acceleration. In particular, this is due the separation assumption between translational and rotation dynamics.

To fulfill the tracking control objective, we then studied the feedback linearizability of the system with respect to $\mathbf{y}^a$ and $\mathbf{y}^b$. Once we proved that $\mathbf{y}^a$ and $\mathbf{y}^b$ are flat output for the considered system, we know that they are also feedback linearizing outputs. Indeed, it exists a feedback control law that linearizes the relation input-output. Then a controller based on the Dynamic Feedback Linearization (DFL) method has been conceived for the tracking of the two sets of outputs, highlighting the related singularities, extended relative degree and the required dynamic extension. Notice that the works in the state of the art considered only the problem of regulating the position of the vehicle (mostly in the 2D case), at most ensuring the positivity of the internal force. On the other hand, in this work we derived a control for the precise tracking of any time-varying desired position and internal force trajectories. Furthermore, we also designed a controller for the precise tracking of a different and never considered output: position plus attitude-related variable.

Finally, for the estimation objective, we showed that if the motion of the moving platform is known, then the state of the system can be estimated using only the onboard IMU and three encoders placed at the other end of the link in order to measure its attitude and length. We then found some nonlinear measurements transformations that allowed us to employ the nonlinear High Gain Observer (presented in Sect. 2.4) to get an estimation of the state and close the control loop preserving the stability of the overall system (see Sect. 4.7).

This theoretical and fundamental results have been extensively validated either experimentally or through numerical simulations. The first implemented hierarchical controller has been validated experimentally and employed for the execution of landing and takeoff on/from a sloped surface as well. The corresponding results are presented in Chap. 5.

The rest of the theoretical results has been instead validated by a deep simulation campaign. We also analyzed the limits of the methods under non ideal conditions such as noisy measurements, parameter uncertainties, etc. Only the dynamic feedback linearizing controller for the output $\mathbf{y}^b$ has not been validated yet. Indeed, we proved the feedback linearizability of the system with respect to $\mathbf{y}^b$ only very recently, without having the possibility to conduct a validating campaign. This will be done in the near future. For the same reason, notice that this results has not been published neither. On the other hand, all the other results related to the control of $\mathbf{y}^a$ and the state estimation problems has been published in [10]. The proof of the differential flatness with respect to $\mathbf{y}^b$ and the design of a hierarchical control to track a desired trajectory of $\mathbf{y}^b$ have been partially presented in [11].

All the previous mentioned results are obviously still valid if there is not a link actuator and the link length is fixed, if the platform is still, and even if the system motion is constrained on a 2D vertical plane. On the other hand, in this particular constrained case, we proved a very interesting result. We showed that only the on-board IMU is needed to retrieve the estimation of the state. Also in this case, we found some non linear measurements and state transformations that bring the system in a canonical form for which the High Gain Observer can be applied. These results have been published in [12, 13].

We also considered another interesting particular case, i.e., when the link actuator is not active but *passive*. In this case the force produced by the link actuator is constant and cannot be controlled. One could made this mechanism with simple springs, like in a retractable leash. This solution simplifies the complexity of the system and reduces the weight as well, thus its could be preferable in some cases, e.g., to make the system easily portable by a human operator. The only controllable actuation is then given by the orientable thrust generated by the underactuated aerial vehicle. Having one controllable actuator less, it appears that the internal force along the link is not controllable anymore, and the attitude angle neither. We proved that only the output $\mathbf{y}^c$, containing the position of the vehicle and its rotation about the thrust axis, is a differentially flat/dynamic feedback linearizable output (see Sects. 4.4.3 and 4.6.3). These results have been published in [14].

All the previous theoretical results constitute the base to solve practical and more application oriented problems. With a good understanding of the system in exam, one can then better exploit its properties for the accomplishment of particular tasks related to a specific applications. For example, considering the challenging problem of landing and takeoff on/from a sloped surface, e.g., in mountains for search and rescue operations, we showed that a tethered aerial vehicle could be very useful to accomplish the task in a very robust and reliable way. In Chap. 6 we theoretically and experimentally show how we exploited the theoretical results, and in particular the ones related to the controllability of $\mathbf{y}^b$, to solve the problem in a more robust way with respect to the case of a free-flying vehicle. The corresponding methods and results have been published in [11, 15]. For a more detailed introduction and state of the art relative to the problem, we refer the reader to Chap. 6.

Finally, starting from the results obtained on control and state estimation of a single tethered aerial vehicle, we performed the same type of rigorous analysis on a multi-robot system (see Chap. 7). Indeed, since the single tethered aerial vehicle is constrained to fly around the anchoring point, the working space (a simple circle in the planar case) can result very limited for some applications, such as pick and place or inspection in cluttered environments. One possible solution is given by adding a second vehicle attached to the first by a second link. In particular, we considered a system composed by two aerial vehicles laying on a vertical 2D plane, where the first one is attached to the ground and to the second one by two links, in the way of forming a chain of two elements. Considering the second robot as an end-effector, the system appears similar to a two-link planar manipulator where the aerial vehicles are the actuators. It is then clear the increased dexterity of the system. Although it might appear not feasible from a practical point of view using cable-links, actually, it could

Table 4.2 Summary of the contributions on tethered aerial vehicles

System	3D environment		Only 2D env.
One link—Active link actuator	General modeling, State estimation with IMU plus three encoders, [10], Sects. 4.3 and 4.7	Differential flatness and DFL w.r.t. $\mathbf{y}^a$, [10], Sect. 4.6.1, and Differential flatness w.r.t. $\mathbf{y}^b$ [11], DFL w.r.t. $\mathbf{y}^b$, Sect. 4.6.2	State estimation with IMU only, [12, 13], Sect. 4.8
One link—Passive link actuator		DFL and differential flatness w.r.t. $\mathbf{y}^c$, Sect. 4.6.3, [14]	
Application to the problem of landing and takeoff on/from a sloped surface	Problem formulation, condition for robust landing, comparison between free-flight and tethered solutions, Chap. 6, [11, 15]		Trajectory planning for robust maneuvers, Sect. 6.5
Multi robot extension: two linked robots in a chain configuration			Differential flatness and DFL w.r.t. $\mathbf{y}_2^a$, Sect. 7.3 [16], state estimation with IMUs plus two encoders, Sect. 7.4, [17]

be easily implemented using bar-links. Furthermore, due to the peculiarity of the system, the control of such a system is a very interesting and challenging theoretical problem. As for the single tethered case, we proved that the output composed by the elevation and the internal force of the two links is differential flat/dynamic feedback linearizable. Thus we designed a nonlinear controller based on the dynamic feedback linearization technique to achieve the independent tracking of the output entries [16]. Furthermore, we proved that the IMU on-board of the two vehicles together with two encoders, one at the base and one on the first vehicle, measuring the relative link angles, are enough to make the state observable. We then found some nonlinear measurement transformations to bring the system in the canonical form that allows to apply the High Gain Observer [17]. The validity of the method has been proven by numerical simulations. For all the related results we refer the reader to Chap. 7.

Table 4.2 schematically gathers all the contributions of this book on the field of tethered aerial vehicles. For each contribution we report the corresponding chapter or section and the papers in which the results have been published.

4.3 Modeling

In this section we provide the model of a generic tethered aerial system consisting of an *unidirectional-thrust aerial vehicle* that is tethered by a generic *link* (e.g., a cable, a rope, a chain, a bar or a strut) whose length can be changed by an *active link*

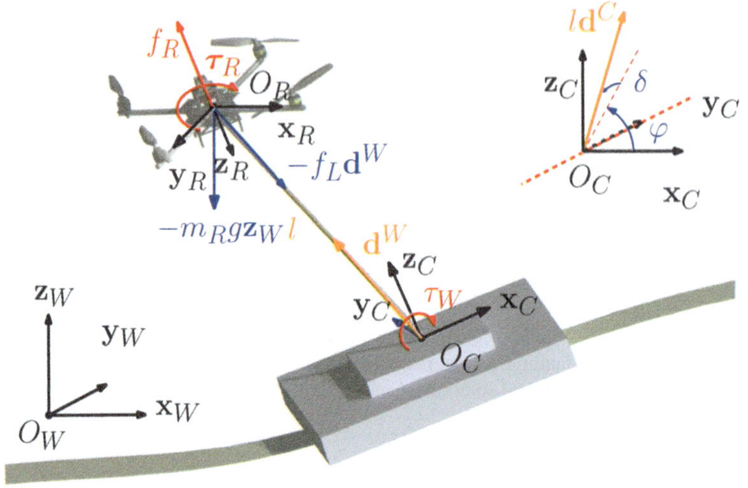

Fig. 4.1 Left: representation of the system and its main variables. Top right corner: parametrization of the unit vector $\mathbf{d}^C$. The red line shows the singularities of the parametrization, avoidable in the planning phase.

actuator. If the link is a cable-like link, then the link actuator is a winch that can roll up or unroll the link. On the other hand, if the link is a telescopic bar-like link, then the link actuator is a prismatic actuator that can extend or shorten the bar. A brief modeling of the link and of the link actuator was presented in Sects. 3.1 and 3.3.2, respectively. Although the bar-link case is not very common, it is still mechanically feasible and allows us to derive a very general model for which we can consider both positive and negative link internal forces (tensions and compressions, respectively). Doing so we expand the validity of the theoretical results. The link actuator is in turn fixed on a *moving platform* by a passive 3D spherical joint. The platform moves generically in the 3D space and can be, e.g., a ground vehicle moving on any kind of terrain, a marine vessel, or even another aerial vehicle. Figure 4.1 depicts the systems and its main definitions.

Consider a fixed *world frame*, $\mathcal{F}_W$ with axes $\{\mathbf{x}_W, \mathbf{y}_W, \mathbf{z}_W\}$ and origin O_W. In particular, $\mathbf{z}_W$ is opposite to the gravity vector. Two body frames, $\mathcal{F}_C$ and $\mathcal{F}_R$ with axes $\{\mathbf{x}_C, \mathbf{y}_C, \mathbf{z}_C\}$ and $\{\mathbf{x}_R, \mathbf{y}_R, \mathbf{z}_R\}$, and origins O_C and O_R, are rigidly attached to the platform and to the aerial vehicle, respectively. The position of O_C in $\mathcal{F}_W$ is described by the vector $\mathbf{p}_C^W = [x_C \ y_C \ z_C]^\top \in \mathbb{R}^3$. Similarly, O_R is set on the CoM of the aerial vehicle, whose position in $\mathcal{F}_W$ is described by the vector $\mathbf{p}_R^W = [x_R \ y_R \ z_R]^\top \in \mathbb{R}^3$.

Although already done in Sect. 3.2, we recall one more time the main modeling assumptions and definitions for a generic aerial vehicle. The aerial vehicle is modeled as a rigid body with mass $m_R \in \mathbb{R}_{>0}$ and positive definite diagonal inertia matrix $\mathbf{J}_R \in \mathbb{R}_{>0}^{3\times3}$ expressed in $\mathcal{F}_R$ and relative to O_R. The angular velocity of $\mathcal{F}_R$ with respect to $\mathcal{F}_W$, expressed in $\mathcal{F}_R$, is denoted by $\boldsymbol{\omega}_R \in \mathbb{R}^3$. The aerial vehicle configuration is fully described by $\mathbf{p}_R^W$ and by the rotation matrix $\mathbf{R}_R \in \mathsf{SO}(3)$, representing the position and orientation of $\mathcal{F}_R$ w.r.t. $\mathcal{F}_W$. The aerial vehicle motion can be controlled acting on four inputs: $f_R \in \mathbb{R}$ and $\boldsymbol{\tau}_R = [\tau_{Rx}\ \tau_{Ry}\ \tau_{Rz}]^\top \in \mathbb{R}^3$, where f_R is the magnitude of the thrust force[1] $\mathbf{f}_R = -f_R\mathbf{z}_R$ applied at O_R and parallel to $\mathbf{z}_R$, and $\boldsymbol{\tau}_R$ is the control moment vector expressed in $\mathcal{F}_R$.

The moving platform configuration is described by $\mathbf{p}_C^W$ and $\mathbf{R}_C \in \mathsf{SO}(3)$, representing the rotation from $\mathcal{F}_C$ to $\mathcal{F}_W$. The angular velocity of $\mathcal{F}_C$ w.r.t. $\mathcal{F}_W$, expressed in $\mathcal{F}_C$, is denoted by $\boldsymbol{\omega}_C \in \mathbb{R}^3$. The platform is an independent sub-system whose motion *(i)* is not influenced by the aerial vehicle dynamics and *(ii)* can only be measured online. In this way, the results can be applied to a broader class of moving platforms including, e.g., human controlled vehicles.

The link connects the aerial vehicle to the moving platform. One end of the link is attached to the aerial vehicle at O_R through a passive 3D spherical joint and the other end is attached to the platform at O_C, through a second passive 3D spherical joint. Having the link directly attached to the CoM of the aerial vehicle allows to decouple the rotational dynamics to the translational one. This assumption is very common in the literature of aerial physical interaction, and is practically easy to meet. Indeed, with a wise mechanical design one can minimize the distance between the CoM and the link attaching point. As explained in Sect. 3.1, we assume negligible link mass and inertia with respect to the ones of the aerial vehicle and negligible deformations and elasticity.

The direction of the link is described by the unit vector[2] $\mathbf{d}^C \in \mathsf{S}^2$ expressed in $\mathcal{F}_C$ thus allowing to express the aerial vehicle position relative to the moving platform.[3] Nevertheless, known $\mathbf{p}_C^W$ and $\mathbf{R}_C$, one can still control $\mathbf{p}_R^W$ by inverse kinematics. The unit vector $\mathbf{d}^C$ can be parametrized with the *elevation angle*, $\varphi \in [0, 2\pi]$, and the *azimuth angle*, $\delta \in [-\frac{\pi}{2}, \frac{\pi}{2}]$, as

$$\mathbf{d}^C = \begin{bmatrix}\cos\delta\cos\varphi & -\sin\delta & \cos\delta\sin\varphi\end{bmatrix}^\top, \tag{4.1}$$

where δ is the angle between $\mathbf{d}^C$ and the vertical plane $\{\mathbf{x}_C, \mathbf{z}_C\}$, whereas φ is the angle between the projection of $\mathbf{d}^C$ on $\{\mathbf{x}_C, \mathbf{z}_C\}$ and $\mathbf{x}_C$, see Fig. 4.1. This particular choice lets the singularity of the parametrization correspond to the points along $\mathbf{y}_C$, whereas the classical spherical parametrization has the singularity along $\mathbf{z}_C$, which corresponds to the common vertical link orientation (when, e.g., the vehicle has to fly above the moving platform).

The link length and the intensity of the internal force are denoted by $l \in \mathbb{R}_{\geq 0}$ and $f_L \in \mathbb{R}_{\geq 0}$, respectively. The link actuator is fixed to the moving platform in the proximity of O_C and is used to control l and f_L in a coordinated action with the

[1] For generality we consider both positive and negative thrust. Although aerial vehicles can usually provide only positive thrust, actually, the variable pitch solution can provide negative thrust as well. However, if only positive thrust is allowed, our controller is still valid, since this constraint can be meet in the planning phase as explained later in the paper.

[2] $\mathsf{S}^2 = \{\mathbf{v} \in \mathbb{R}^3 \mid \|\mathbf{v}\| = 1\}$.

[3] We express the link orientation in $\mathcal{F}_C$ because the goal is to control the aerial vehicle position relative to the moving platform rather than to $\mathcal{F}_W$.

aerial vehicle thrust force. We recall from Sect. 3.3.2 that the link actuator exerts an input torque $\tau_W \in \mathbb{R}$ that controls the link length. The constant rotational inertia and constant radius of the link actuator are denoted by $J_W \in \mathbb{R}_{>0}$ and $r_W \in \mathbb{R}_{>0}$, respectively.

Since the link is attached to the aerial vehicle center of mass by a passive rotational joint, the aerial vehicle rotational dynamics is independent from the translational dynamics and it is equal to the one derived in Sect. 3.2:

$$\dot{\mathbf{R}}_R = \mathbf{R}_R \boldsymbol{\Omega}_R \tag{4.2}$$

$$\mathbf{J}_R \dot{\boldsymbol{\omega}}_R = \mathbf{J}_R \boldsymbol{\omega}_R \times \boldsymbol{\omega}_R + \boldsymbol{\tau}_R. \tag{4.3}$$

The linear velocity of the aerial vehicle is obtained differentiating $\mathbf{p}_R^W = \mathbf{p}_C^W + l\mathbf{R}_C\mathbf{d}^C$:

$$\dot{\mathbf{p}}_R^W = \mathbf{R}_C \left(\dot{\mathbf{p}}_C^C + l\boldsymbol{\Omega}_C\mathbf{d}^C + \dot{l}\mathbf{d}^C + l\dot{\mathbf{d}}^C \right). \tag{4.4}$$

To derive the dynamic equations of the generalized coordinates $\mathbf{q} = [l \;\; \varphi \;\; \delta]^\top$ we use the Newton-Euler approach,[4] solving the balance of the forces acting on O_R in $\mathcal{F}_W$, and the balance of momenta about the axis of the link actuator (see Sect. 3.3.2):

$$m_R \ddot{\mathbf{p}}_R^W = -f_L \mathbf{R}_C \mathbf{d}^C - f_R \mathbf{z}_R^W - m_R g \mathbf{z}_W^W \tag{4.5}$$

$$\bar{J}_W \ddot{l} = \bar{\tau}_W + f_L, \tag{4.6}$$

where $\bar{J}_W = J_W / r_W^2$, $\bar{\tau}_W = \tau_W / r_W$. The acceleration $\ddot{\mathbf{p}}_R^W$ is obtained by further differentiating (4.4), i.e.,

$$\ddot{\mathbf{p}}_R^W = \mathbf{R}_C [\bar{\mathbf{a}}_x + \mathbf{J}_\mathrm{q} \ddot{\mathbf{q}}], \tag{4.7}$$

where $\bar{\mathbf{a}}_x = \boldsymbol{\Omega}_C(\dot{\mathbf{p}}_C^C + l\boldsymbol{\Omega}_C\mathbf{d}^C + 2\mathbf{J}_\mathrm{q}\dot{\mathbf{q}}) + \ddot{\mathbf{p}}_C^C + l\dot{\boldsymbol{\Omega}}_C\mathbf{d}^C + \dot{\mathbf{J}}_\mathrm{q}\dot{\mathbf{q}}$ and

$$\mathbf{J}_\mathrm{q} = \begin{bmatrix} \cos\delta\cos\varphi & -l\cos\delta\sin\varphi & -l\cos\varphi\sin\delta \\ -\sin\delta & 0 & -l\cos\delta \\ \cos\delta\sin\varphi & l\cos\delta\cos\varphi & -l\sin\delta\sin\varphi \end{bmatrix}. \tag{4.8}$$

Replacing (4.7) into (4.5) and after some algebra we get

$$m_R \mathbf{J}_\mathrm{q} \ddot{\mathbf{q}} + f_L \mathbf{d}^C = \underbrace{-m_R \bar{\mathbf{a}}_x}_{\mathbf{a}_x} \underbrace{- m_R g \mathbf{R}_C^\top \mathbf{e}_3}_{\mathbf{a}_g} \underbrace{- f_R \mathbf{R}_C^\top \mathbf{R}_R \mathbf{e}_3}_{\mathbf{a}_{f_R}}, \tag{4.9}$$

[4]Since one of our goal is the one of controlling the internal force along the link, it results helpful to have its analytical expression. Newton-Euler is then the proper method to use as explained in Sect. 2.1. Indeed, using the Lagrangian formalism, we would not have obtained the sought internal force expression.

Finally, gathering (4.9) and (4.6) we obtain a square system

$$\underbrace{\begin{bmatrix} m_R \mathbf{J_q} & \mathbf{d}^C \\ \mathbf{J}_{W\mathbf{q}} & -1 \end{bmatrix}}_{\mathbf{W}} \begin{bmatrix} \ddot{\mathbf{q}} \\ f_L \end{bmatrix} = \underbrace{\begin{bmatrix} -\mathbf{a}_x - \mathbf{a}_g \\ 0 \end{bmatrix}}_{\mathbf{a}} + \underbrace{\begin{bmatrix} -\mathbf{a}_{f_R} \\ \bar{\tau}_W \end{bmatrix}}_{\mathbf{a_u}}, \tag{4.10}$$

where $\mathbf{J}_{W\mathbf{q}} = [\bar{J}_W \;\; 0 \;\; 0] \in \mathbb{R}^{1\times 3}$ and $\mathbf{W} \in \mathbb{R}^{4\times 4}$ is invertible if and only if $l \neq 0$ and $\delta \neq \pm\pi/2$, that correspond to the singularities of the pseudo-spherical coordinates of O_R. Inverting equation (4.10) out of these singularities, we obtain

$$\ddot{\mathbf{q}} = \begin{bmatrix} \mathbf{I}_3 \; \mathbf{0}_{3\times 1} \end{bmatrix} \mathbf{W}^{-1}(\mathbf{a} + \mathbf{a_u}) = \sigma(\mathbf{x}, \mathbf{X}_C^2, \mathbf{u}) \tag{4.11}$$

$$f_L = \begin{bmatrix} \mathbf{0}_{1\times 3} \; 1 \end{bmatrix} \mathbf{W}^{-1}(\mathbf{a} + \mathbf{a_u}), \tag{4.12}$$

where $\mathbf{x} = (\mathbf{q}, \dot{\mathbf{q}}, \mathbf{R}_R, \boldsymbol{\omega}_R)$ is the system state, $\mathbf{u} = [f_R \;\; \boldsymbol{\tau}_R^\top \;\; \bar{\tau}_W]^\top = [u_1 \;\; u_2 \;\; u_3 \;\; u_4 \;\; u_5]^\top$ is the vector of inputs, and

$$\mathbf{X}_C^j = (\mathbf{x}_C^0, \mathbf{x}_C^1, \ldots, \mathbf{x}_C^j) \text{ for } j \in \mathbb{N}_{>0}, \tag{4.13}$$

with

$$\begin{aligned} \mathbf{x}_C^i &= (\mathbf{v}_C^{C(i-1)}, \boldsymbol{\omega}_C^{(i-1)}) \text{ for } i = 1, 2, \ldots \\ \mathbf{x}_C^0 &= (\mathbf{p}_C^W, \mathbf{R}_C) \end{aligned} \tag{4.14}$$

where $\mathbf{v}_C^C = \mathbf{R}_C^\top \frac{d\mathbf{p}_C}{dt}$. $\mathbf{X}_C^j$ in (4.13) gathers the terms related to the motion of the platform.

Gathering Eqs. (4.2), (4.3) and (4.11) we have a complete description of the system dynamics:

$$\ddot{\mathbf{q}} = \begin{bmatrix} \mathbf{I}_3 \; \mathbf{0}_{3\times 1} \end{bmatrix} \mathbf{W}^{-1}(\mathbf{a} + \mathbf{a_u}) = \sigma(\mathbf{x}, \mathbf{X}_C^2, \mathbf{u}) \tag{4.15a}$$

$$\dot{\mathbf{R}}_R = \mathbf{R}_R \boldsymbol{\Omega}_R \tag{4.15b}$$

$$\mathbf{J}_R \dot{\boldsymbol{\omega}}_R = \mathbf{J}_R \boldsymbol{\omega}_R \times \boldsymbol{\omega}_R + \boldsymbol{\tau}_R. \tag{4.15c}$$

In the following we show some results that apply to simpler and yet relevant cases of the generic system considered so far. It is then useful to show how we can particularize (4.15) for those cases:

- *Fixed platform*: one can simply set $\mathbf{x}_C^i = (\mathbf{p}_C^{C(i)}, \boldsymbol{\omega}_C^{(i-1)}) = (\mathbf{0}, \mathbf{0})$ for all $i \geq 1$.
- *Fixed link length*: this represents the case in which there is no link actuator and the aerial vehicle is tethered directly to the platform by a link with a constant length, l. The model can be easily adapted considering all the time derivatives of l equal to zero, i.e., $\dot{l} = \ddot{l} = \ldots = 0$. In this case l becomes a parameter and the generalized coordinates reduce to $\mathbf{q} = [\varphi \;\; \delta]^\top$. The dynamics can be easily derived from (4.10)

considering only the first three row and $\ddot{l} = 0$. The last row is always verified since $\bar{\tau}_W$ represents the reaction force of the anchoring point.

- *Passive link actuator*: The link actuator is considered passive because a non-controllable constant torque $\tau'_W \in \mathbb{R}$ is applied along the longitudinal axis of the cylinder ($\tau_W(t) = \tau'_W$ for all $t \in \mathbb{R}_{\geq 0}$), e.g., generated by a simple constant torque spring. The length l can be then controlled only by the action of the thrust provided by the aerial vehicle. The choice of a passive link actuator instead of a controllable one makes the system simpler and easily portable by a human operator. On the other hand, as it will be clear in Sect. 4.6.3, the price to pay will be a reduced control authority on the variables of the system. In particular, the internal force of the link cannot be regulated to an arbitrary value while following a position trajectory. However it can be maintained within a desired bound, if the desired trajectory is well planned. For the sake of studying the feedback linearizability it is convenient to rewrite the model of the system (4.15) in the following Lagrangian form:

$$\mathbf{M}\ddot{\mathbf{q}} + \mathbf{c} + \mathbf{g} + \mathbf{n} + \mathbf{w} = \mathbf{Q}\mathbf{u}_R, \tag{4.16}$$

where, $\mathbf{u}_R = [u_1 \;\; \mathbf{u}_2^\top]^\top = [f_R \;\; \boldsymbol{\tau}_R^\top]^\top \in \mathbb{R}^4$ is the reduce input vector, $\mathbf{M}(\mathbf{q}) \in \mathbb{R}^{3\times3}$ is the positive definite inertia matrix, $\mathbf{c}(\mathbf{q}, \dot{\mathbf{q}}, \dot{\mathbf{p}}_C^C, \omega_C) \in \mathbb{R}^3$ contains all the centrifugal/Coriolis terms, $\mathbf{g}(\mathbf{q}, \mathbf{R}_C) \in \mathbb{R}^3$ contains all the gravity terms, $\mathbf{n}(\mathbf{q}, \ddot{\mathbf{p}}_C^C, \dot{\omega}_C) \in \mathbb{R}^3$ contains the terms depending on the acceleration of the moving platform, $\mathbf{w}(\tau_W) \in \mathbb{R}^3$ contains the terms depending on the constant torque winch, and $\mathbf{Q}(\mathbf{q}, \mathbf{R}_R, \mathbf{R}_C) \in \mathbb{R}^{3\times4}$ is related to the generalized forces, referred as $\mathbf{a}_\mathbf{u}$, performing work on $\mathbf{q}$, such that

$$\mathbf{a}_\mathbf{u} = \mathbf{Q}(\mathbf{q}, \mathbf{R}_R, \mathbf{R}_C)\mathbf{u} = [-\mathbf{J}_\mathbf{q}^\top \mathbf{R}_C^\top \mathbf{R}_R \mathbf{e}_3 \;\; \mathbf{0}_{3\times3}]\mathbf{u}_R, \tag{4.17}$$

For sake of brevity we do not report here the full expression of each term. Notice that in Lagrangian representation of the dynamics the internal force of the link does not appear, differently from (4.11) and (4.12). This is useful since, as it will be clear in Sect. 4.6.3, the internal force along the link is not part of the differential flat/feedback linearizing output for the tethered system with passive link actuator.

- *Reduced model*: in the following Sect. 4.8 we shall show some particular results proven only in the particular case in which the link length and the ground platform are fixed, and the vehicle is restricted to move on a 2D vertical plane (for simplicity let us consider the plane $(\mathbf{x}_W, \mathbf{z}_W)$). To write the dynamics of the system in this particular conditions starting from the previous model, we first parametrize the vehicle attitude by the *Euler-angles* roll, pitch and yaw, (ϕ, θ, ψ), being the angles of rotation about the major axes $(\mathbf{x}_R, \mathbf{y}_R, \mathbf{z}_R)$. We then impose $\phi, \psi = 0$ that implies $\mathbf{y}_R = \mathbf{y}_W$. Furthermore we can consider the azimuth constantly to zero, i.e., $\delta = \dot{\delta} = \ddot{\delta} = 0$. Under this conditions the system degrees of freedom reduce to two, described by the new vector $\mathbf{q}' = [\varphi \;\; \theta]^\top \in \mathbb{R}^2$, actuated by the total thrust and the torque about $\mathbf{y}_R$, $\mathbf{u}' = [u_1 \;\; u_3]^\top = [f_R \;\; \tau_{R_y}]^\top \in \mathbb{R}^2$. Under those assumptions and definitions, we can rewrite (4.15) as:

Table 4.3 Particular cases of the general tethered aerial system and corresponding dynamics

Particular cases	Dynamics
Static platform	(4.15) with $\mathbf{x}_C^i = (\mathbf{p}_C^{C(i)}, \boldsymbol{\omega}_C^{(i-1)}) = (\mathbf{0}, \mathbf{0})$ for all $i \geq 1$
No link actuator	(4.10) with $\dot{l} = \ddot{l} = \ldots = 0$ and $\bar{\tau}_W = f_L$
Passive link actuator	(4.15) with $\tau_W = \tau_W'$ constantly
Reduced model	(4.18) with $\mathbf{q}' = [\varphi\ \theta]^\top$ and $\mathbf{u}' = [u_1\ u_3]^\top = [f_R\ \tau_{Ry}]^\top$

$$\mathbf{M}'(\mathbf{q}')\ddot{\mathbf{q}}' + \mathbf{g}'(\mathbf{q}') = \mathbf{Q}'(\mathbf{q}')\mathbf{u}', \tag{4.18}$$

where

$$\mathbf{M}' = \begin{bmatrix} m_R l^2 & 0 \\ 0 & J_{R22} \end{bmatrix}, \quad \mathbf{g}' = \begin{bmatrix} m_R l g \mathbf{d}^\perp \cdot \mathbf{e}_3 \\ 0 \end{bmatrix},$$
$$\mathbf{Q}' = \begin{bmatrix} -l\mathbf{R}_R\mathbf{e}_3 \cdot \mathbf{d}^\perp & 0 \\ 0 & 1 \end{bmatrix}, \quad \mathbf{u}' = \begin{bmatrix} f_R \\ \tau_{Ry} \end{bmatrix} = \begin{bmatrix} u_1 \\ u_3 \end{bmatrix},$$

and J_{Rkm} with $k, m \in \{1, 2, 3\}$ corresponds to the element of the matrix $\mathbf{J}_R$ in position k, m, $\mathbf{d} = [\cos(\varphi)\ 0\ \sin(\varphi)]^\top$, $\mathbf{d}^\perp = [-\sin(\varphi)\ 0\ \cos(\varphi)]^\top$.

Table 4.3 gathers all the previous particular cases and the corresponding dynamics.

4.4 Differential Flatness

We recall from Sect. 2.2 that a system is *differentially flat* when it exists (at least) an output, called *flat output*, such that the states and the inputs can be expressed as an algebraic function of the flat output and its derivatives, up to a finite order [18]. Thus the flatness property would fulfill our first objective, i.e., compute analytically and offline the nominal state and input of the system required to exactly track a desired output trajectory. In fact, this property is commonly used for control to compute the feed-forward terms, and for planning and optimization to generate feasible trajectories, especially for nonholonomic and underactuated systems, considering the input limitations. Furthermore, the differential flatness property tells a lot about which are the independently controllable outputs and which is the required degree of smoothness for the corresponding desired trajectories.

For an unidirectional-thrust aerial vehicle, it is well known that the position of its CoM, $\mathbf{p}_R$, and the rotation along the $\mathbf{z}_R$ axis, ψ, (called yaw angle when using the Euler-angles parametrization, see Sect. 2.1.3) are differentially flat outputs[5] [19, 20].

If we tether the aerial vehicle to a fixed point by a link with constant length, the link constraints the vehicle to fly on a sphere. Intuitively, the vehicle can still control the position on the sphere (two d.o.f.) and the yaw angle, but not the position along the longitudinal axis of the link. Indeed, every force component applied along this axis will not produce motion due to the kinematic constraint. On the other hand, it will stretch or compress the link, according to its direction. In other words, the vehicle cannot change the distance from the anchoring point but it can control the intensity of the internal force along the link. If then the link length is actuated, the link actuator can control its length and in thus the distance of the vehicle from the anchoring point. This intuition tell us that the output $\mathbf{y}^a = [y_1^a \ y_2^a \ y_3^a \ y_4^a \ y_5^a]^\top = [l \ \varphi \ \delta \ f_L \ \eta_i]^\top \in \mathbb{R}^5$, where η_i is a more "generalized" yaw angle (see the following) is a differentially flat output. We shall prove this result in Sect. 4.4.1.

Furthermore, let us consider for simplicity the system constrained to move on a 2D vertical plane. It is easy to notice that the link internal force and the angle between the thrust and the horizon, defined by $\vartheta_A \in \mathbb{R}$ (see later for the formal definition), are directly connected. Indeed the higher the internal force, the more the thrust tends to be parallel to the link axis, i.e., ϑ_A tends to φ. This second intuition makes us believe that the output $\mathbf{y}^b = [y_1^b \ y_2^b \ y_3^b \ y_4^b \ y_5^b]^\top = [l \ \varphi \ \delta \ \vartheta_A \ \eta_i]^\top \in \mathbb{R}^5$ is a differentially flat output as well. We shall prove this result in Sect. 4.4.2.

To prove the differential flatness we show how to express $\mathbf{x}$ and $\mathbf{u}$ as function of $\mathbf{y}^a$ or $\mathbf{y}^b$ and some of their relative derivatives, in the standard case in which η_i is the yaw angle. We recall that the state consists of the parametrization of $\mathbf{d}^C$, its velocity, the attitude of the aerial vehicle and its angular velocity, while the inputs are the thrust, the total torque provided by the robot and the winch torque. As we already said, we suppose $\mathbf{X}_C^0$ and its derivatives known (see (4.13)).

4.4.1 Stress-Related Flat Output

Let us define $\mathbf{y}_1^a = [y_1^a \ y_2^a \ y_3^a]^\top = [l \ \varphi \ \delta]^\top$ such that $\mathbf{y}^a = [\mathbf{y}_1^{a\top} \ y_4^a \ y_5^a]^\top = [\mathbf{q}^\top \ f_L \ \eta_i]^\top$. We have directly that

$$\mathbf{q} = \mathbf{y}_1^a, \quad \dot{\mathbf{q}} = \dot{\mathbf{y}}_1^a. \tag{4.19}$$

Then, from (4.6) we obtain the expression of $\bar{\tau}_W$

$$\bar{\tau}_W = \bar{J}_W \ddot{y}_1^a - y_4^a = f_1(\ddot{y}_1^a, y_4^a). \tag{4.20}$$

In order to find the expression of the missing states and inputs, one can notice that from (4.7) the linear acceleration of the aerial vehicle can be written as function of

[5] And thus also dynamic feedback linearizing outputs.

$\mathbf{y}_1^a, \dot{\mathbf{y}}_1^a, \ddot{\mathbf{y}}_1^a$, the linear velocity of the platform, its rotation and their time derivatives, i.e., $\ddot{\mathbf{p}}_R^W = \ddot{\mathbf{p}}_R^W(\mathbf{y}_1^a, \dot{\mathbf{y}}_1^a, \ddot{\mathbf{y}}_1^a, \mathbf{X}_C^2)$. Then, from (4.5), we can write the thrust vector as function of only the outputs, their derivatives and known quantities related to the trajectory of the moving platform, indeed:

$$\begin{aligned} f_R \mathbf{R}_R \mathbf{e}_3 &= -m_R \ddot{\mathbf{p}}_R^W(\mathbf{y}_1^a, \dot{\mathbf{y}}_1^a, \ddot{\mathbf{y}}_1^a, \mathbf{X}_C^2) - y_4 \mathbf{R}_C \mathbf{d}^C(\mathbf{y}_1) - m_R g \mathbf{e}_3 \\ &= \mathbf{f}_2(\mathbf{y}_1^a, \dot{\mathbf{y}}_1^a, \ddot{\mathbf{y}}_1^a, y_4, \mathbf{X}_C^2). \end{aligned} \tag{4.21}$$

Then, exploiting y_5, one can apply the same method presented in [20] in order to obtain the attitude of the aerial vehicle, its angular velocity and the total provided torque. For the sake of brevity we omit here the full re-derivation of all these functions.

Proposition *The model (4.15), is differentially flat with respect to the flat output* $\mathbf{y}^a = [l \;\; \varphi \;\; \delta \;\; f_L \;\; \eta_i]^\top$ *where* η_i *is the yaw angle of* $\mathbf{R}_R$. *In other words, the state and the inputs can be written as algebraic function of* $\mathbf{y}^a$ *and a finite number of its derivatives:*

$$\mathbf{x} = \mathbf{f}_\mathbf{x}^a(\mathbf{y}^a, \ldots, \mathbf{y}^{a\,(4)}, \mathbf{X}_C^2) \tag{4.22}$$

$$\mathbf{u} = \mathbf{f}_\mathbf{u}^a(\mathbf{y}^a, \ldots, \mathbf{y}^{a\,(4)}, \mathbf{X}_C^2). \tag{4.23}$$

4.4.2 Attitude-Related Flat Output

Firstly, we have to properly define ϑ_A in the 3D environment. Let us define a new reference frame, $\mathcal{F}_L$ centered in $\mathbf{p}_C$ and with axes $\{\mathbf{x}_L, \mathbf{y}_L, \mathbf{z}_L\}$ such that:

$$\mathbf{z}_L = \mathbf{z}_W, \quad \mathbf{y}_L = \frac{\mathbf{z}_L \times \mathbf{d}}{\|\mathbf{z}_L \times \mathbf{d}\|}, \quad \mathbf{x}_L = \mathbf{y}_L \times \mathbf{z}_L. \tag{4.24}$$

If $\mathbf{d}$ is parallel to $\mathbf{z}_W$, one can choose any arbitrary $\mathbf{y}_L$. For example, if the aerial vehicle is moving such that $\mathbf{d}$ is parallel to $\mathbf{z}_W$ at a certain time t_i, during that instant we can simply keep $\mathbf{y}_L$ constant. Practically, starting from a non singular condition, one would define a certain threshold $\epsilon \in \mathbb{R}_{>0}$ such that if $\|\mathbf{z}_W \times \mathbf{d}\| < \epsilon$, $\mathbf{y}_L$ is kept constant. In particular we can define the vertical plane $\mathcal{P}_L$ that includes $\mathbf{d}$, i.e., $\mathcal{P}_L = \{\mathbf{v} \in \mathbb{R}^3 \mid \exists\, \lambda_1, \lambda_2 \in \mathbb{R} \,:\, \mathbf{v} = \lambda_1 \mathbf{x}_L + \lambda_2 \mathbf{z}_L\}$; and the projection of $-\mathbf{z}_R$ into $\mathcal{P}_L$ defined by:

$$\boldsymbol{\alpha} = [\alpha_1 \;\; \alpha_2 \;\; \alpha_3]^\top = -\mathbf{P}_L \mathbf{z}_R, \tag{4.25}$$

where $\mathbf{P}_L = [\mathbf{x}_L \;\; \mathbf{0} \;\; \mathbf{z}_L]^\top \in \mathbb{R}^{3\times 3}$ is the projector on $\mathcal{P}_L$, with respect to $\mathcal{F}_L$. Finally we define ϑ_A as the angle between $\boldsymbol{\alpha}$ and $\mathbf{z}_L$:

$$\vartheta_A = \operatorname{atan2}(\alpha_1, \alpha_3) = \operatorname{atan2}\left(-\mathbf{e}_1^\top \mathbf{P}_L \mathbf{z}_R, -\mathbf{e}_3^\top \mathbf{P}_L \mathbf{z}_R\right). \tag{4.26}$$

Figure 4.2 shows how ϑ_A is graphically defined.

In order to demonstrate that $\mathbf{y}^b$ is also a flat output we show that there exists a bijective map between $\mathbf{y}^b$ and $\mathbf{y}^a$. Considering $\mathbf{X}_C^2$ (see 4.13) as a known input, the map from $\mathbf{y}^a$ to $\mathbf{y}^b$ and their derivatives, i.e., $(\mathbf{y}^b, \ldots, \mathbf{y}^{b(4)}) = \mathbf{g}_b(\mathbf{y}^a, \ldots, \mathbf{y}^{a(4)}, \mathbf{X}_C^2)$ is simply given by the flatness of the system w.r.t. $\mathbf{y}^a$. In fact, given $\mathbf{y}^a$ and its derivatives, one can compute the nominal state and inputs with (4.22) and (4.23). Then, from Eq. (4.26), it is easy to compute $\mathbf{y}^b$ and the relative derivatives.

Regarding the opposite sense of the map, i.e., from $\mathbf{y}^b$ to $\mathbf{y}^a$:

$$(\mathbf{y}^a, \ldots, \mathbf{y}^{a(4)}) = \mathbf{g}_a(\mathbf{y}^b, \ldots, \mathbf{y}^{b(4)}, \mathbf{X}_C^2), \tag{4.27}$$

the map is immediate for $\mathbf{q}$ and ψ. Let us define $\mathbf{y}_1^b = [y_1^b\ y_2^b\ y_3^b]^\top = [l\ \varphi\ \delta]^\top$ such that $\mathbf{y}^b = [\mathbf{y}_1^{b\top}\ y_4^b\ y_5^b]^\top = [\mathbf{q}^\top\ \vartheta_A\ \eta_i]^\top$. We have that $\mathbf{y}_1^a = \mathbf{y}_1^b$ and $y_5^a = y_5^b$. Then we can retrieve f_L, and so y_4^a, from $\mathbf{y}^b$ and its derivatives projecting both sides of (4.5) on the plane $\mathcal{P}_L$. Not considering the second equations (always zero), after some algebra we can obtain:

$$y_4^a = f_L = -m_R[0\ 1]^\top \mathbf{T}^{-1}[\mathbf{e}_1\ \mathbf{e}_3]^\top \mathbf{P}_L(\ddot{\mathbf{p}}_R + g\mathbf{z}_W), \tag{4.28}$$

where $\mathbf{T}(\mathbf{y}^b) = [\mathbf{e}_1\ \mathbf{e}_3]^\top[\mathbf{P}_L\mathbf{z}_R\ \ \mathbf{P}_L\mathbf{d}]$ is invertible if and only if $\mathbf{z}_R \not\perp \mathcal{P}_L$ and $\mathbf{P}_L\mathbf{z}_R \nparallel \mathbf{P}_L\mathbf{d}$. Finally, to retrieve the derivatives of f_L one can simply differentiate (4.28) w.r.t. time. This proves that between $\mathbf{y}^a$ and $\mathbf{y}^b$, and their derivatives, there is a bijective map.

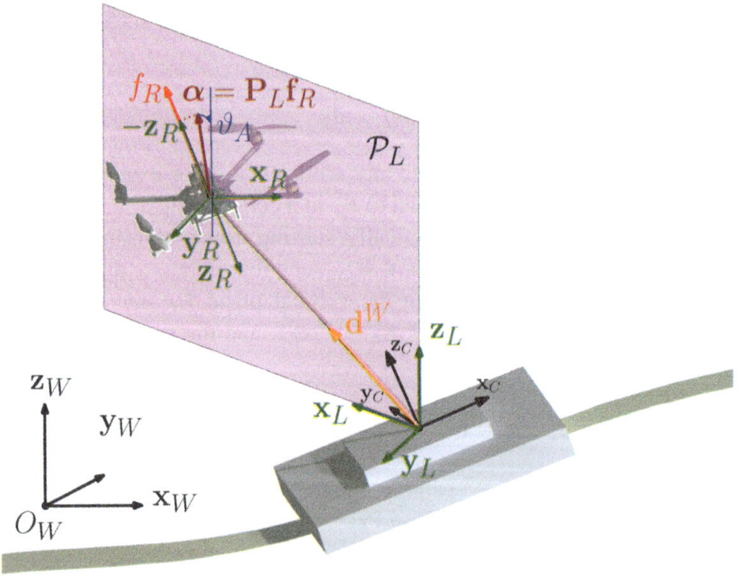

Fig. 4.2 Visual description of the angle ϑ_A

Combining (4.22), (4.23) and (4.27), state and inputs of the system can be written as an algebraic function of $\mathbf{y}^b$ its derivatives, and the known quantity $\mathbf{X}_C^2$, proving that $\mathbf{y}^b$ is a flat output:

$$\mathbf{x} = \mathbf{f}_\mathbf{x}^a(\mathbf{g}_a(\mathbf{y}^b, \ldots, \mathbf{y}^{b(4)}, \mathbf{X}_C^2), \mathbf{X}_C^2) = \mathbf{f}_\mathbf{x}^b(\mathbf{y}^b, \ldots, \mathbf{y}^{b(4)}, \mathbf{X}_C^2) \tag{4.29}$$

$$\mathbf{u} = \mathbf{f}_\mathbf{u}^a(\mathbf{g}_a(\mathbf{y}^b, \ldots, \mathbf{y}^{b(4)}, \mathbf{X}_C^2), \mathbf{X}_C^2) = \mathbf{f}_\mathbf{u}^b(\mathbf{y}^b, \ldots, \mathbf{y}^{b(4)}, \mathbf{X}_C^2). \tag{4.30}$$

Proposition *The model (4.15), is differentially flat with respect to the flat output* $\mathbf{y}^b = [l \ \varphi \ \delta \ \vartheta_A \ \eta_i]^\top$ *where* η_i *is the yaw angle of* $\mathbf{R}_R$. *In other words, the state and the inputs can be written as algebraic function of* $\mathbf{y}^b$ *and a finite number of its derivatives, i.e., Eqs. (4.29) and (4.30).* □

4.4.3 Differential Flatness for Passive Link Actuator

Let us define $\mathbf{y}^c = [\mathbf{y}_1^{c\top} \ y_4^c]^\top = [\mathbf{q}^\top \ \eta_i]^\top \in \mathbb{R}^4$. Part of the state is directly given by $\mathbf{y}_1^c$, i.e., $\mathbf{q} = \mathbf{y}_1^c$, $\dot{\mathbf{q}} = \dot{\mathbf{y}}_1^c$. Only the part of the state related to the rotational dynamics and the inputs have still to be derived. From (4.6) we can write the internal force as function of $\ddot{\mathbf{y}}_1^c$:

$$f_L = \bar{J}_W \ddot{l} - \bar{\tau}_W = f_L(\ddot{\mathbf{y}}_1^c). \tag{4.31}$$

Then, it is exactly the same procedure as in Sect. 4.4.1. One can write $\mathbf{y}_1^a = \mathbf{y}_1^c$, $y_4^a = f_L(\ddot{\mathbf{y}}_1^c)$, $y_5^a = y_5^c$, and then use (4.22) and (4.23) to compute nominal state and input.

Proposition *The model (4.2), (4.3) and (4.16), is differentially flat with respect to the flat output* $\mathbf{y}^c = [l \ \varphi \ \delta \ \eta_i]^\top$ *where* η_i *is the yaw angle of* $\mathbf{R}_R$. *In other words the state and the inputs can be written as algebraic function of* $\mathbf{y}^c$ *and a finite number of its derivatives.*

$$\mathbf{x} = \mathbf{f}_\mathbf{x}^c(\mathbf{y}^c, \ldots, \mathbf{y}^{c(4)}, \mathbf{X}_C^2) \tag{4.32}$$

$$\mathbf{u} = \mathbf{f}_\mathbf{u}^c(\mathbf{y}^c, \ldots, \mathbf{y}^{c(4)}, \mathbf{X}_C^2). \tag{4.33}$$

4.4.3.1 Discussion on Link Internal Force Regulation

We already remarked that, differently from the active link actuator case, the internal force along the link is not part of the flat output for this system. This means that its value cannot be directly controlled. On the contrary, from (4.31), it is a byproduct of the desired output trajectory, and in particular of the desired link length acceleration. Nevertheless, in order to keep the internal force always within a desired bound $\mathcal{B}_{f_L} = [\underline{f_L}, \overline{f_L}]$ where $\underline{f_L}, \overline{f_L} \in \mathbb{R}$, we can exploit the flatness of the system to design suitable desired trajectories of $\mathbf{y}^c$. In particular, from Eq. (4.31) we have that $f_L \in \mathcal{B}_{f_L}$ if and only if $\ddot{l} \in \mathcal{B}_{\ddot{l}} = [\underline{\ddot{l}}, \overline{\ddot{l}}]$ where $\overline{\ddot{l}} = (\bar{\tau}'_W + \overline{f_L})/\bar{J}_W$ and $\underline{\ddot{l}} = (\bar{\tau}'_W + \underline{f_L})/\bar{J}_W$. In other words the constraint on the internal force can be translated using the flatness into a constraint on the desired trajectory of l.

Notice that the steady configuration, $\ddot{l} = 0$, belongs to $\mathcal{B}_{\ddot{l}}$ if and only if $\underline{\ddot{l}} \leq 0 \leq \overline{\ddot{l}}$ that in turn means $-\overline{f_L} \leq \bar{\tau}'_W \leq -\underline{f_L}$. In particular, if for $\ddot{l} = 0$ we want a desired internal force value $f_L{}^\star \in \mathcal{B}_{f_L}$, we have to design the passive link actuator such that $\bar{\tau}'_W = \tau'_W / r_W = -f_L{}^\star$. Another parameter of the link actuator that can be optimized is its inertia $\bar{J}_W$. Indeed it affects how $\mathcal{B}_{f_L}$ is mapped on $\mathcal{B}_{\ddot{l}}$, e.g., if we make $\bar{J}_W$ small enough, big variations of $\ddot{l}$ imply small variations of the internal force and thus an almost constant internal force, $f_L \approx f_L{}^\star$.

4.5 Hierarchical Control

In this section we exploit the previously proven flatness in order to design a controller based on a hierarchical method. This is very common for the control of aerial vehicles because it allows to separate the control of the attitude from the one of the translational dynamics. This is convenient especially for commercial available platforms for which one cannot directly control the spinning velocity of the propellers. Often, one can only control the angular velocity or the attitude of the vehicle, sending the desired Euler angles. Then, internally to the platform, a closed low level controller track the desired references.

To cope with these problems, we propose a simpler control strategy, based on hierarchical techniques, that can be easily implemented in every platform. Indeed, thanks to the separation between outer loop (normally position control) and inner loop (attitude control) controls, one can easily adapt the proposed controller to the specific platform functionalities. Similar techniques were successfully implemented and tested in, e.g., [5, 6], to only stabilize the position of the vehicle. However, those methods cannot be directly applied to solve our problem because they are designed for different systems, although similar.

The validity of the method has been proven experimentally. The related results are shown in Sect. 5.2.

4.5.1 Force-Related Hierarchical Control

In the following we design a hierarchical controller for controlling the output $\mathbf{y}^a$, namely the position of the aerial vehicle and the internal force along the link. The controller is based on the cascaded structure between the translational and rotational dynamics.

Given a desired position trajectory $\mathbf{p}_R^C(t)$, defined in terms of the generalized coordinates $\mathbf{y}_1^{ad}(t) = \mathbf{q}(t)^d$ we define

$$\ddot{\mathbf{q}}^\star = \ddot{\mathbf{q}}^d + \mathbf{K}_\mathbf{q}^D(\dot{\mathbf{q}}^d - \dot{\mathbf{q}}) + \mathbf{K}_\mathbf{q}^P(\dot{\mathbf{q}}^d - \dot{\mathbf{q}}), \tag{4.34}$$

where $\mathbf{K}_\mathbf{q}^P, \mathbf{K}_\mathbf{q}^D \in \mathbb{R}_{>0}^{3\times3}$ are diagonal matrices. The vector $\ddot{\mathbf{q}}^\star$ could be seen as the desired acceleration that lets $\mathbf{q}$ follow the desired trajectory $\mathbf{q}^d(t)$ using a PD strategy. In case of model uncertainties or disturbances one can add an integral term as well.

Then, given a desired trajectory for the internal force of the link $y_4^a = f_L(t)^d$, and inverting the balance of momenta on the link actuator (4.6), we compute the link actuator torque as

$$\bar{\tau}_W = \bar{J}_W \ddot{l}^\star - f_L{}^d. \tag{4.35}$$

To finally implement $\ddot{\mathbf{q}}^\star$ we compute the desired thrust vector inverting the balance force equation on O_R (4.5), like for the flatness computation in (4.21)

$$\mathbf{f}_R^\star = f_R \mathbf{R}_R^\star \mathbf{e}_3 = \mathbf{R}_C(-\mathbf{a}_x - \mathbf{a}_g - f_L{}^d \mathbf{d}^C - m_R \mathbf{J}_\mathbf{q} \ddot{\mathbf{q}}^\star). \tag{4.36}$$

From the desired thrust vector we derive the input f_R as

$$f_R = \left\| -\mathbf{a}_x - \mathbf{a}_g - f_L{}^d \mathbf{d}^C - m_R \mathbf{J}_\mathbf{q} \ddot{\mathbf{q}}^\star \right\|, \tag{4.37}$$

and the desired z-axis of $\mathcal{F}_R$, i.e., $\mathbf{z}_R^\star = \mathbf{R}_R^\star \mathbf{e}_3 = \mathbf{f}_R^\star / f_R$.

The desired yaw angle ψ^d together with $\mathbf{z}_R^\star$ define the desired attitude of the vehicle described by $\mathbf{R}_R^\star$. In fact, given ψ^d we define $\mathbf{x}'_R = \mathbf{R}_\mathbf{z}(\psi^d)\mathbf{e}_1$ where $\mathbf{R}_\mathbf{z}(\psi^d)$ is the rotation matrix describing the rotation of ψ^d around $\mathbf{z}_W$. The axis $\mathbf{x}'_R$ represents the desired heading of the aerial vehicle. The desired attitude is computed creating an orthonormal basis using the vectors $\mathbf{x}'_R$ and $\mathbf{z}_R^\star$ that is given by $\mathbf{R}_R^\star = [\mathbf{x}_R^\star \; \mathbf{y}_R^\star \; \mathbf{z}_R^\star]$ where,

$$\mathbf{y}_R^\star = \frac{\mathbf{z}_R^\star \times \mathbf{x}'_R}{\left\|\mathbf{z}_R^\star \times \mathbf{x}'_R\right\|}, \qquad \mathbf{x}_R^\star = \frac{\mathbf{y}_R^\star \times \mathbf{z}_{R\star}}{\left\|\mathbf{y}_R^\star \times \mathbf{z}_{R\star}\right\|}. \tag{4.38}$$

This concludes the design of the outer loop control. Given the tracking error it computes the desired link actuator torque $\bar{\tau}_W$, the desired thrust intensity f_R and the desired attitude $\mathbf{R}_R^\star$.

If the considered platform accept as input the desired attitude and thrust, we can simply send the previous quantities as control commands. If instead the platform is controlled in thrust/angular velocity or thrust/torque, we shall show the design of an inner loop control that computes the desired angular velocity, $\boldsymbol{\omega}_R^\star$, or the desired torque, $\boldsymbol{\tau}_R$, respectively, in order to track the desired attitude computed by the outer loop control.

Let us define the attitude error [21] by the vector $\mathbf{e}_R \in \mathbb{R}^3$, computed as

$$[\mathbf{e}_R]_\times = -\frac{1}{2}(\mathbf{R}_R^{\star\top}\mathbf{R}_R - \mathbf{R}_R^\top\mathbf{R}_R^\star), \tag{4.39}$$

where $[\mathbf{e}_R]_\times$ corresponds to the skew symmetric matrix relative to $\mathbf{e}_R$. In order to steer $\mathbf{e}_R$ to zero, if the vehicle is controlled in angular velocity, we compute the desired angular velocity based on a P controller,

$$\boldsymbol{\omega}_R^\star = \boldsymbol{\omega}_R^d + \mathbf{K}_\omega^P \mathbf{e}_R, \tag{4.40}$$

where $\mathbf{K}_\omega^P \in \mathbb{R}_{>0}^{3\times3}$ is a diagonal matrix. If instead the vehicle is controlled in torque we first define the desired angular acceleration based on a PD controller,

$$\dot{\boldsymbol{\omega}}_R^\star = \dot{\boldsymbol{\omega}}_R^d + \mathbf{K}_\omega^D(\boldsymbol{\omega}_R^d - \boldsymbol{\omega}_R) + \mathbf{K}_\omega^P \mathbf{e}_R, \tag{4.41}$$

where $\mathbf{K}_\omega^D \in \mathbb{R}_{>0}^{3\times3}$ is a diagonal matrix, and $\boldsymbol{\omega}_R^d$ and $\dot{\boldsymbol{\omega}}_R^d$ are the nominal angular velocity and acceleration, respectively, computed by the flatness from the desired output trajectory as in Sect. 4.4.1. Inverting the rotational dynamics we can finally find the input torque $\boldsymbol{\tau}_R$,

$$\boldsymbol{\tau}_R = -\mathbf{J}_R\boldsymbol{\omega}_R \times \boldsymbol{\omega}_R + \mathbf{J}_R\dot{\boldsymbol{\omega}}_R^\star. \tag{4.42}$$

If the inner loop is sufficiently faster than the outer loop, the asymptotic convergence of $\mathbf{q}$ to $\mathbf{q}^d$ is guaranteed. Figure 4.3a shows a schematic representation of the controller $\mathbf{u} = \Gamma_{\mathrm{HC}}^a(\mathbf{x}, \mathbf{X}_C^2, \mathbf{y}^{ad}(t))$ given by (4.35), (4.37) and (4.42).

Notice that it is easy to rewrite the controller Γ_{HC}^a for the particular cases mentioned in Sect. 4.3. For the case of a passive link actuator one can use the same controller once the nominal internal force is computed by (4.31).

4.5.2 *Attitude-Related Hierarchical Control*

For the control of $\mathbf{y}^b$ we exploit the bijection function between $\mathbf{y}^b$ and $\mathbf{y}^a$, and the previously presented hierarchical controller for the tracking of $\mathbf{y}^a$. In particular, from the desired trajectory $\mathbf{y}^{bd}(t)$, we can compute the equivalent trajectory of $\mathbf{y}^a$, i.e., $(\mathbf{y}^{ad}, \ldots, \mathbf{y}^{ad(4)}) = \mathbf{g}_a(\mathbf{y}^{bd}, \ldots, \mathbf{y}^{bd(4)}, \mathbf{X}_C^2)$. Then we apply the hierarchical controller shown before to effectively track $\mathbf{y}^{ad}(t)$. The hierarchical controller to track $\mathbf{y}^{bd}(t)$ defined by Γ_{HC}^b, is equal to:

$$\mathbf{u} = \Gamma_{\mathrm{HC}}^b(\mathbf{x}, \mathbf{X}_C^2, \mathbf{y}^{bd}(t)) = \Gamma_{\mathrm{HC}}^a(\mathbf{x}, \mathbf{X}_C^2, \mathbf{g}_a(\mathbf{y}^{bd}(t), \mathbf{X}_C^2)). \tag{4.43}$$

Thanks to the fact that $\mathbf{g}_a$ and $\mathbf{g}_b$ are bijective maps, if the closed loop system is able to track $\mathbf{y}^{ad}(t)$, it will implicitly track $\mathbf{y}^{bd}(t)$. Figure 4.3b shows a schematic representation of the controller.

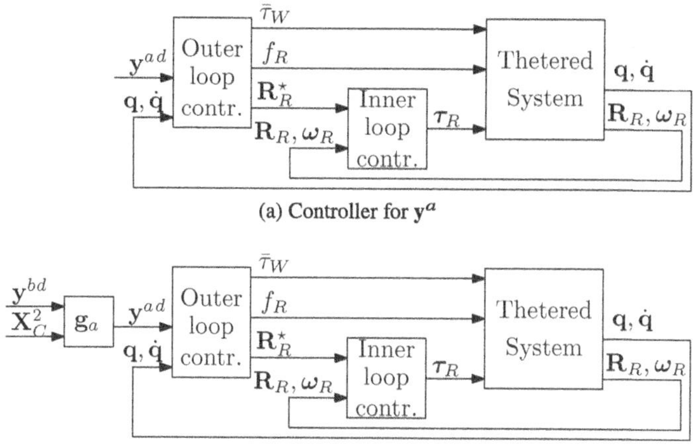

(a) Controller for $\mathbf{y}^a$

(b) Controller for $\mathbf{y}^b$

Fig. 4.3 Schematic representation of the two hierarchical controllers

4.6 Dynamic Feedback Linearization

From the theory we know that a flat output is also an exactly dynamical feedback linearizing output on an open and dense set of the state space [22]. In this section we shall design a controller based on the dynamic feedback linearization technique the problem of the exact tracking of the outputs $\mathbf{y}^a$, $\mathbf{y}^b$, and $\mathbf{y}^c$. For the problem of the exact tracking of such output we will design a controller based on the dynamic feedback linearization technique. We recall that in the following $\mathbf{x} = (\mathbf{q}, \dot{\mathbf{q}}, \mathbf{R}_R, \boldsymbol{\omega}_R)$ is the system state, $\mathbf{u} = [f_R \ \boldsymbol{\tau}_R^\top \ \bar{\tau}_W]^\top = [u_1 \ u_2 \ u_3 \ u_4 \ u_5]^\top$ is the input vector and the outputs are

(i) $\mathbf{y}^a = [y_1^a \ y_2^a \ y_3^a \ y_4^a \ y_5^a]^\top = [l \ \varphi \ \delta \ f_L \ \eta_i]^\top \in \mathbb{R}^5$,
(ii) $\mathbf{y}^b = [y_1^b \ y_2^b \ y_3^b \ y_4^b \ y_5^b]^\top = [l \ \varphi \ \delta \ \vartheta_A \ \eta_i]^\top \in \mathbb{R}^5$ and
(iii) $\mathbf{y}^c = [y_1^c \ y_2^c \ y_3^c \ y_4^c]^\top = [l \ \varphi \ \delta \ \eta_i]^\top \in \mathbb{R}^4$.

4.6.1 *Force-Related Feedback Linearizing Output*

In the previous sections we considered η_i as the yaw angle when the rotation of $\mathcal{F}_R$ is parametrized by the Euler angles. However, in the following we consider any generic parametrization $\boldsymbol{\eta} = [\eta_1 \ \eta_3 \ \eta_3]^\top \in \mathbb{R}^3$ of $\mathbf{R}_R$, such that $\mathbf{R}_R = \mathbf{R}_R(\boldsymbol{\eta})$ and $\dot{\boldsymbol{\eta}} = \mathbf{T}_\eta \boldsymbol{\omega}_R$ where $\mathbf{T}_\eta(\boldsymbol{\eta}) \in \mathbb{R}^{3\times3}$ is given by the particular parametrization [23]. From (4.3) the dynamics of $\boldsymbol{\eta}$ is

$$\ddot{\boldsymbol{\eta}} = \underbrace{\dot{\mathbf{T}}_{\eta}\boldsymbol{\omega}_R + \mathbf{T}_{\eta}\mathbf{J}_R^{-1}(\mathbf{J}_R\boldsymbol{\omega}_R \times \boldsymbol{\omega}_R)}_{\mathbf{b}_{\eta}(\boldsymbol{\eta},\dot{\boldsymbol{\eta}})} + \underbrace{[\,\mathbf{0}_{3\times1} \;\; \mathbf{T}_{\eta}\mathbf{J}_R^{-1} \;\; \mathbf{0}_{3\times1}\,]}_{\mathbf{E}_{\eta}(\boldsymbol{\eta})}\mathbf{u}. \tag{4.44}$$

Then we consider η_i as any entry of $\boldsymbol{\eta}$ such that, in the domain of interest, it holds

$$e_{\eta_i} = \frac{\partial \ddot{\eta}_i}{\partial \tau_{Rz}} = \mathbf{e}_i^\top \mathbf{T}_{\eta}\mathbf{J}_R^{-1}\mathbf{e}_3 \neq 0. \tag{4.45}$$

For example, taking $\boldsymbol{\eta} = [\phi \;\; \theta \;\; \psi]^\top$ as the classical Roll-Pitch-Yaw parametrization of $\mathbf{R}_R$ and $\eta_i = \psi$, we have that

$$\mathbf{T}_{\eta}(\boldsymbol{\eta}) = \begin{bmatrix} 1 & \sin\phi\tan\theta & \cos\phi\tan\theta \\ 0 & \cos\phi & -\sin\phi \\ 0 & \sin\phi\sec\theta & \cos\phi\sec\theta \end{bmatrix} \text{ and } e_{\eta_i} = \frac{1}{J_{R33}}\cos\phi\sec\theta, \tag{4.46}$$

Notice that for this choice (4.45) holds always except for $\phi = \pi/2$ and $\theta = \pi/2$.

Intuitively, only $\bar{\tau}_W$ and $\mathbf{f}_R$ play a role in the control of l, φ, δ and f_L (see (4.10)) and they are not affected by τ_{Rz}. Indeed $\mathbf{f}_R$ is not influenced by rotations along $\mathbf{z}_R$ and therefore not even by the torque τ_{Rz} acting about it. Then it is necessary to complete the set of outputs with a quantity dynamically dependent on τ_{Rz} to have a well-posed tracking problem. Thus, we recall that the output of interest is $\mathbf{y}^a = [y_1^a \;\; y_2^a \;\; y_3^a \;\; y_4^a \;\; y_5^a]^\top = [l \;\; \varphi \;\; \delta \;\; f_L \;\; \eta_i]^\top \in \mathbb{R}^5$.

Applying the feedback linearization technique (see Sect. 2.3), and recalling Eqs. (4.10) and (4.44), we immediately see that (y_1^a, y_2^a, y_3^a) have to be differentiated twice until f_R and $\bar{\tau}_W$ appear. Also y_5^a has to be differentiated twice until $\boldsymbol{\tau}_R$ appears, while y_4^a directly depends on f_R and $\bar{\tau}_W$. Defining $\bar{\mathbf{y}}_1^a = [\ddot{y}_1^a \;\; \ddot{y}_2^a \;\; \ddot{y}_3^a \;\; y_4^a]^\top$ and rearranging (4.10) and (4.44), we can write

$$\begin{bmatrix} \bar{\mathbf{y}}_1^a \\ \ddot{y}_5^a \end{bmatrix} = \begin{bmatrix} \bar{\mathbf{W}}\mathbf{a} \\ b_{\eta_i} \end{bmatrix} + \begin{bmatrix} \bar{\mathbf{W}}\mathbf{U} \\ \mathbf{e}_{\eta_i} \end{bmatrix}\mathbf{u} = \mathbf{b}(\mathbf{x}, \mathbf{X}_C^2) + \mathbf{E}(\mathbf{x}, \mathbf{X}_C^0)\mathbf{u}, \tag{4.47}$$

where $\bar{\mathbf{W}} = \mathbf{W}^{-1}$, $b_{\eta_i} = \mathbf{e}_i^\top \mathbf{b}_{\eta}$, $\mathbf{e}_{\eta_i} = \mathbf{e}_i^\top \mathbf{E}_{\eta}$, the vector $\mathbf{b}(\mathbf{x}, \mathbf{X}_C^2)$ gathers all the terms that do not depend on the inputs and

$$\mathbf{U} = \begin{bmatrix} -\mathbf{R}_C^\top \mathbf{R}_R \mathbf{e}_3 & \mathbf{0}_{3\times3} & \mathbf{0}_{3\times1} \\ 0 & \mathbf{0}_{1\times3} & 1 \end{bmatrix}_{4\times5} \tag{4.48}$$

$$\mathbf{E} = \begin{bmatrix} \bar{\mathbf{W}} & \mathbf{0}_{4\times1} \\ \mathbf{0}_{1\times4} & 1 \end{bmatrix} \begin{bmatrix} -\mathbf{R}_C^\top \mathbf{R}_R \mathbf{e}_3 & \mathbf{0}_{3\times3} & \mathbf{0}_{3\times1} \\ 0 & \mathbf{0}_{1\times3} & 1 \\ 0 & \mathbf{e}_i^\top \mathbf{T}_{\eta}\mathbf{J}_R^{-1} & 0 \end{bmatrix}_{5\times5}. \tag{4.49}$$

We recall that Rearranging the rows of the decoupling matrix $\mathbf{E}$ one can notice that it is clearly singular because $\boldsymbol{\tau}_R$ does not appear in the expression of $\bar{\mathbf{y}}_1^a$.

As explained in Sect. 2.3, to obtain a full rank matrix we insert a dynamic compensator considering as new input $\bar{\mathbf{u}} = [\ddot{u}_1 \;\; u_2 \;\; u_3 \;\; u_4 \;\; \ddot{u}_5]^\top$, where $\ddot{u}_1$ and $\ddot{u}_5$ are the second derivative of f_R and $\bar{\tau}_W$, respectively. Under this definition $\bar{\mathbf{y}}_1^a$ and y_5^a have to be differentiated twice to see the new inputs appear:

$$\begin{bmatrix} \ddot{\bar{\mathbf{y}}}_1^a \\ \ddot{y}_5^a \end{bmatrix} = \begin{bmatrix} \ddot{\bar{\mathbf{W}}}(\mathbf{a}+\mathbf{a_u}) + 2\dot{\bar{\mathbf{W}}}(\dot{\mathbf{a}}+\dot{\mathbf{a}}_\mathbf{u}) + \bar{\mathbf{W}}(\ddot{\mathbf{a}}+\ddot{\mathbf{a}}_\mathbf{u}) \\ b_i + \mathbf{e}_{\eta_i}\bar{\mathbf{u}} \end{bmatrix}, \tag{4.50}$$

where $\ddot{\mathbf{a}}_\mathbf{u}$, after replacing the system dynamics, results:

$$\ddot{\mathbf{a}}_\mathbf{u} = \bar{\ddot{\mathbf{a}}}_\mathbf{u} + \begin{bmatrix} -u_1\mathbf{R}_C^\top\mathbf{R}_R\left[\mathbf{J}_R^{-1}\boldsymbol{\tau}_R\right]_\times \mathbf{e}_3 - \ddot{u}_1\mathbf{R}_C^\top\mathbf{R}_R\mathbf{e}_3 \\ \ddot{u}_5 \end{bmatrix}. \tag{4.51}$$

Since $\mathbf{J}_R$ is diagonal, i.e., $J_{Rkm} = 0$ for $k \neq m$ and $k, m \in \{1, 2, 3\}$, writing the skew symmetric matrix relative to $\mathbf{J}_R^{-1}\boldsymbol{\tau}_R$ and doing some algebra we obtain

$$\left[\mathbf{J}_R^{-1}\boldsymbol{\tau}_R\right]_\times \mathbf{e}_3 = \left[-\frac{\mathbf{e}_2}{J_{R11}}\ \frac{\mathbf{e}_1}{J_{R22}}\ \mathbf{0}_{3\times1}\right]\boldsymbol{\tau}_R. \tag{4.52}$$

Replacing Eqs. (4.52) and (4.51) into (4.50) we obtain

$$\begin{bmatrix} \ddot{\bar{\mathbf{y}}}_1^a \\ \ddot{y}_5^a \end{bmatrix} = \underbrace{\begin{bmatrix} \ddot{\bar{\mathbf{W}}}(\mathbf{a}+\mathbf{a_u}) + 2\dot{\bar{\mathbf{W}}}(\dot{\mathbf{a}}+\dot{\mathbf{a}}_\mathbf{u}) + \bar{\mathbf{W}}(\ddot{\mathbf{a}}+\bar{\ddot{\mathbf{a}}}_\mathbf{u}) \\ b_i \end{bmatrix}}_{\bar{\mathbf{b}}(\bar{\mathbf{x}},\mathbf{X}_C^4)} + \underbrace{\begin{bmatrix} \bar{\mathbf{W}}\bar{\mathbf{U}} \\ \mathbf{e}_{\eta_i} \end{bmatrix}}_{\bar{\mathbf{E}}(\bar{\mathbf{x}},\mathbf{X}_C^0)} \bar{\mathbf{u}}, \tag{4.53}$$

where $\bar{\mathbf{U}} = \begin{bmatrix} -\mathbf{R}_C^\top\mathbf{R}_R\mathbf{T} & \mathbf{0}_{3\times1} \\ \mathbf{0}_{1\times4} & 1 \end{bmatrix}$, $\bar{\mathbf{x}} = (\mathbf{q}, \dot{\mathbf{q}}, \mathbf{R}_R, \boldsymbol{\omega}_R, f_R, \dot{f}_R, \bar{\tau}_W, \dot{\bar{\tau}}_W)$ is the extended state, and $\mathbf{T} = [\mathbf{e}_3\ \ -\frac{u_1}{J_{R11}}\mathbf{e}_2\ \ \frac{u_1}{J_{R22}}\mathbf{e}_1\ \ \mathbf{0}_{3\times1}] \in \mathbb{R}^{3\times4}$. Changing the order of the inputs as in $\tilde{\mathbf{u}} = [u_1\ u_2\ u_3\ u_5\ u_4]^\top$, the decoupling matrix becomes $\tilde{\mathbf{E}} = \begin{bmatrix} \tilde{\mathbf{E}}_1 & \mathbf{0}_{1\times3} \\ \tilde{\mathbf{e}}_3 & e_{\eta_i} \end{bmatrix}$, where

$\tilde{\mathbf{E}}_1 = \bar{\mathbf{W}}\begin{bmatrix} \tilde{\mathbf{U}}_1 & \mathbf{0}_{3\times1} \\ \mathbf{0}_{1\times3} & 1 \end{bmatrix}$, $\tilde{\mathbf{e}}_3 = \left[0\ \mathbf{e}_i^\top\mathbf{T}_\eta\mathbf{J}_R^{-1}\mathbf{e}_1\ \mathbf{e}_i^\top\mathbf{T}_\eta\mathbf{J}_R^{-1}\mathbf{e}_2\ 0\right]$,

$\tilde{\mathbf{U}}_1 = -\mathbf{R}_C^\top\mathbf{R}_R\left[\mathbf{e}_3\ -\frac{u_1}{J_{R11}}\mathbf{e}_2\ \frac{u_1}{J_{R22}}\mathbf{e}_1\right] = -\mathbf{R}_C^\top\mathbf{R}_R\tilde{\mathbf{T}}$.

The original decoupling matrix $\bar{\mathbf{E}}$ is invertible if $\tilde{\mathbf{E}}$ is invertible, or equivalently, due to its canonical form, if $\tilde{\mathbf{E}}_1$ is invertible and e_{η_i} is nonzero. Since the matrices $\mathbf{R}_C$, $\mathbf{R}_R$ and $\bar{\mathbf{W}}$ are always full rank (except in the model singularities, i.e., $l = 0$ and $\delta = \pm\pi/2$), then $\tilde{\mathbf{U}}_1$ is invertible whenever $\tilde{\mathbf{T}}$ is full rank, i.e., if $u_1 \neq 0$, indeed $\det(\tilde{\mathbf{T}}) = u_1^2/(J_{R11}J_{R22})$.

In the cases in which the thrust u_1 is not zero and with the opportune parametrization of $\mathbf{R}_R$, using the control law

$$\bar{\mathbf{u}} = \bar{\mathbf{E}}(\bar{\mathbf{x}},\mathbf{X}_C^0)^{-1}\left[-\bar{\mathbf{b}}(\bar{\mathbf{x}},\mathbf{X}_C^4) + \mathbf{v}^a\right], \tag{4.54}$$

where $\mathbf{v}^a = [v_1^a\ v_2^a\ v_3^a\ v_4^a\ v_5^a]^\top \in \mathbb{R}^5$ are virtual inputs, we obtain

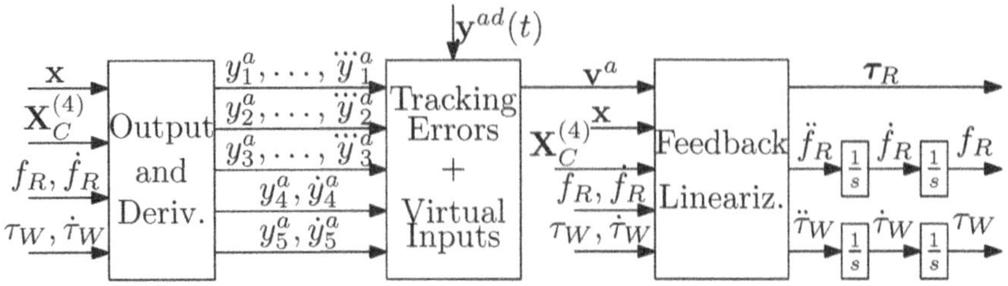

Fig. 4.4 Block diagram representation of the control strategy. © 2020 IEEE. Reprinted, with permission, from [10]

$$y_1^{a(4)} = v_1^a, \quad y_2^{a(4)} = v_2^a, \quad y_3^{a(4)} = v_3^a, \quad y_4^{a(2)} = v_4^a, \quad y_5^{a(2)} = v_5^a. \tag{4.55}$$

Furthermore, the total relative degree with respect to $\mathbf{y}^a$ is $r^a = 16$ that corresponds to the dimension of the extended state $\bar{\mathbf{x}}$ that is $\bar{n} = 16$. Indeed it is composed by $\mathbf{x}$ (of dimension 12) plus the four states of the dynamic compensator. Therefore the system is exactly dynamic feedback linearizable and the linearized system (4.55) does not have an internal dynamics [24].

4.6.1.1 Outer Linear Controller

In the following we will omit the subscript $\bullet^a$ since the same outer-loop control can be applied to track both $\mathbf{y}^a$ and $\mathbf{y}^b$, after the application of the opportune linearizing control law.

The tracking of any given desired trajectory, $y_i^d(t) \in C^3$ for $i = 1, 2, 3$ and $y_j^d(t) \in C^1$ for $j = 4, 5$ can be achieved applying any linear control technique to the equivalent linear system (4.55), as depicted in Fig. 4.4. As explained in Sect. 2.3, it is sufficient to use as outer loop a simple controller based on the pole placing technique. Setting the virtual control inputs as

$$v_i = y_i^{d(4)} + \mathbf{k}_i^\top \boldsymbol{\xi}_i, \qquad v_j = y_j^{d(2)} + \mathbf{k}_j^\top \boldsymbol{\xi}_j, \tag{4.56}$$

where $\boldsymbol{\xi}_i = [\xi_i^{(3)} \ \ddot{\xi}_i \ \dot{\xi}_i \ \xi_i]^\top \in \mathbb{R}^4, \boldsymbol{\xi}_j = [\dot{\xi}_j \ \xi_j]^\top \in \mathbb{R}^2, \xi_i = y_i^d - y_i$ and $\xi_j = y_j^d - y_j$ are the tracking errors, one can set the poles of the error dynamics through the gains $\mathbf{k}_i \in \mathbb{R}^4_{>0}$ and $\mathbf{k}_j \in \mathbb{R}^2_{>0}$, for $i = 1, 2, 3$ and $j = 4, 5$, to obtain a sufficiently fast exponentially tracking of the desired trajectories.

All the previous results are summarized in the following

Proposition *For the analyzed system it exists at least one parametrization* $\boldsymbol{\eta}$ *of* $\mathbf{R}_R$ *and one of its elements* η_i *such that* $\mathbf{y}^a = [l \ \varphi \ \delta \ f_L \ \eta_i]^\top$ *is an exact feedback linearizing output for each state, except if* $l = 0$, $\delta = \pm\pi/2$ *and* $u_1 = 0$ *(zero thrust case). Furthermore, considering as input* $\bar{\mathbf{u}} = [\ddot{u}_1 \ u_2 \ u_3 \ u_4 \ \ddot{u}_5]^\top$*, the control law*

$\bar{\mathbf{u}} = \Gamma^a_{DFL}(\bar{\mathbf{x}}, \mathbf{X}^4_C, \mathbf{y}^{ad}(t))$ *defined by (4.54) and (4.56) exponentially steers* $\mathbf{y}^a$ *along the desired trajectories* $y^{ad}_i(t) \in C^3$ *for i=1, 2, 3, and* $y^{ad}_j(t) \in C^1$ *for j=4, 5.* □

Remark In order to implement the exact tracking control laws (4.54)–(4.56) the only needed quantities are

- the desired output trajectory and its derivatives $y^{\star d}_i, \dot{y}^{\star d}_i, \ddot{y}^{\star d}_i, \dddot{y}^{\star d}_i, \ddddot{y}^{\star d}_i$ for $i = 1, 2, 3$, and $y^{\star d}_j, \dot{y}^{\star d}_j, \ddot{y}^{d}_j$ for $j = 4, 5$, where $y^\star_i$ is either y^a_i or y^b_i
- a measurement of system state, i.e., $\mathbf{x} = (\mathbf{q}, \dot{\mathbf{q}}, \mathbf{R}_R, \boldsymbol{\omega}_R)$
- the internal state of the compensators $f_R, \dot{f}_R, \bar{\tau}_W, \dot{\bar{\tau}}_W$
- the position and orientation of the moving platform and their derivatives $\mathbf{X}^4_C$. □

We recall that an explicit measurement of the output and its derivatives is not needed at all, since they are algebraic functions of the state and input.

4.6.2 Attitude-Related Feedback Linearizing Output

Now we are ready to show that $\mathbf{y}^b = [y^b_1\ y^b_2\ y^b_3\ y^b_4\ y^b_5]^\top = [l\ \varphi\ \delta\ \vartheta_A\ \eta_i]^\top \in \mathbb{R}^5$ is a feedback linearizing output. Similarly to Sect. 4.4.2, we firstly show that $\mathbf{y}^b$ and its derivatives can be written as a function of $\mathbf{y}^a$ and its derivatives, and $\mathbf{X}^i_C$ with $i > 0$. In particular, we shall show that $[\mathbf{y}^{b(4)}_1\ \ddot{y}^b_4\ \ddot{y}^b_5]^\top$, linearly depends on $[\mathbf{y}^{a(4)}_1\ \ddot{y}^a_4\ \ddot{y}^a_5]^\top$, i.e., there exists the functions $\mathbf{b}_{\mathbf{y}^b}(\mathbf{y}^a_1, \ldots, \mathbf{y}^{a(3)}_1, \mathbf{X}^4_C)$ and $\mathbf{E}_{\mathbf{y}^b}(\mathbf{y}^a_1, \mathbf{X}^0_C)$ such that

$$\begin{bmatrix} \mathbf{y}^{b(4)}_1 \\ \ddot{y}^b_4 \\ \ddot{y}^b_5 \end{bmatrix} = \mathbf{b}_{\mathbf{y}^b}(\mathbf{y}^a_1, \ldots, \mathbf{y}^{a(3)}_1, \mathbf{X}^4_C) + \mathbf{E}_{\mathbf{y}^b}(\mathbf{y}^a_1, \mathbf{X}^0_C) \begin{bmatrix} \mathbf{y}^{a(4)}_1 \\ \ddot{y}^a_4 \\ \ddot{y}^a_5 \end{bmatrix}. \tag{4.57}$$

Let us first extract $\mathbf{f}_R$ from (4.9), and differentiate it twice, showing the dependence on $\mathbf{y}^a$ and its derivatives:

$$\mathbf{f}_R = \underbrace{-\mathbf{R}_C(\mathbf{a}_x + \mathbf{a}_g)}_{\mathbf{b}'_{\mathbf{f}_R}(\mathbf{y}^a_1, \dot{\mathbf{y}}^a_1, \mathbf{X}^2_C)} + \underbrace{[m_R \mathbf{R}_C \mathbf{J}_\mathbf{q}\ \mathbf{d}]}_{\mathbf{E}_{\mathbf{f}_R}(\mathbf{y}^a_1, \mathbf{X}^0_C) = [\mathbf{E}'_{\mathbf{f}_R} \mathbf{e}''_{\mathbf{f}_R}]} \bar{\mathbf{y}}^a \tag{4.58}$$

$$\ddot{\mathbf{f}}_R = \mathbf{b}_{\mathbf{f}_R}(\mathbf{y}^a_1, \ldots, \mathbf{y}^{a(3)}_1, \mathbf{X}^4_C) + \mathbf{E}_{\mathbf{f}_R}(\mathbf{y}^a_1, \mathbf{X}^0_C)\ddot{\bar{\mathbf{y}}}^a, \tag{4.59}$$

where $\mathbf{b}_{\mathbf{f}_R}$ gathers all the terms that do not depend on $\ddot{\bar{\mathbf{y}}}^a$. Similarly, can write $\mathbf{z}_R$ and its derivatives as function of $\mathbf{f}_R$ and its derivatives, and thus as function of $\mathbf{y}^a$ and its derivatives:

$$\mathbf{z}_R = \mathbf{f}_R / \|\mathbf{f}_R\| \tag{4.60}$$

$$\dot{\mathbf{z}}_R = 1/\|\mathbf{f}_R\|(\mathbf{I}_3 - \mathbf{z}_R\mathbf{z}_R^\top)\dot{\mathbf{f}}_R = \mathbf{E}_{\mathbf{z}_R}(\mathbf{y}^a_1, \dot{\mathbf{y}}^a_1, \mathbf{X}^2_C)\dot{\mathbf{f}}_R \tag{4.61}$$

$$\ddot{\mathbf{z}}_R = \mathbf{b}_{\mathbf{z}_R}(\mathbf{y}^a_1, \ldots, \mathbf{y}^{a(3)}_1, \mathbf{X}^4_C) + \mathbf{E}_{\mathbf{z}_R}\mathbf{E}_{\mathbf{f}_R}\ddot{\bar{\mathbf{y}}}^a, \tag{4.62}$$

where $\mathbf{b}_{\mathbf{z}_R}$ gathers all the terms that do not depend on $\ddot{\mathbf{y}}^a$.

Let us now compute the second derivative of $\boldsymbol{\alpha}$, expressing it as function of $\mathbf{y}^a$ and its derivatives, replacing the previous equations where necessary.

$$\ddot{\boldsymbol{\alpha}} = \mathbf{b}_{\boldsymbol{\alpha}}(\mathbf{y}_1^a, \dots, \mathbf{y}_1^{a(3)}, \mathbf{X}_C^4) - \mathbf{P}_L \mathbf{E}_{\mathbf{z}_R} \mathbf{E}_{\mathbf{f}_R} \ddot{\mathbf{y}}^a, \tag{4.63}$$

where $\mathbf{b}_{\boldsymbol{\alpha}}$ gathers all the terms that do not depend on $\ddot{\mathbf{y}}^a$.

Finally, we can compute the derivatives of $y_4^b = \vartheta_A$ in order to obtain the form in (4.57)

$$\dot{y}_4^b = \frac{1}{\alpha_1^2 + \alpha_3^2} \begin{bmatrix} -\alpha_3 & \alpha_1 \end{bmatrix} \begin{bmatrix} \dot{\alpha}_1 \\ \dot{\alpha}_3 \end{bmatrix} = \underbrace{\frac{1}{\|\boldsymbol{\alpha}\|^2} \boldsymbol{\alpha}^\top \begin{bmatrix} -\mathbf{e}_3 & \mathbf{0} & \mathbf{e}_1 \end{bmatrix}}_{\mathbf{e}_{y_4^b}(\mathbf{y}_1^a, \dots, \mathbf{y}_1^{a(3)}, \mathbf{X}_C^4)} \dot{\boldsymbol{\alpha}} \tag{4.64}$$

$$\ddot{y}_4^b = \mathbf{b}'_{y_4^b} - \mathbf{e}_{y_4^b} \ddot{\boldsymbol{\alpha}} = \mathbf{b}_{y_4^b} - \mathbf{e}_{y_4^b} \mathbf{P}_L \mathbf{E}_{\mathbf{z}_R} \mathbf{E}_{\mathbf{f}_R} \ddot{\mathbf{y}}^a \tag{4.65}$$

where $\mathbf{b}'_{y_4^b}$ and $\mathbf{b}_{y_4^b}$ gather all the terms that do not depend on $\ddot{\boldsymbol{\alpha}}$ and $\ddot{\mathbf{y}}^a$, respectively.

Finally, noticing that $\mathbf{y}_1^b = \mathbf{y}_1^a$ and $y_5^b = y_5^a$ we can write the following equation, equal to the sought form (4.57)

$$\begin{bmatrix} \mathbf{y}_1^{b(4)} \\ \ddot{y}_4^b \\ \ddot{y}_5^b \end{bmatrix} = \underbrace{\begin{bmatrix} \mathbf{0} \\ \mathbf{b}_{y_4^b} \\ 0 \end{bmatrix}}_{\mathbf{b}_{\mathbf{y}^b}(\mathbf{y}_1^a, \dots, \mathbf{y}_1^{a(3)}, \mathbf{X}_C^4)} + \underbrace{\begin{bmatrix} \mathbf{I}_3 & 0 & 0 \\ \mathbf{e}'_{y_4^b} & \mathbf{e}''_{y_4^b} & 0 \\ \mathbf{0} & 0 & 1 \end{bmatrix}}_{\mathbf{E}_{\mathbf{y}^b}(\mathbf{y}_1^a, \mathbf{X}_C^0)} \begin{bmatrix} \mathbf{y}_1^{a(4)} \\ \ddot{y}_4^a \\ \ddot{y}_5^a \end{bmatrix}, \tag{4.66}$$

where $\mathbf{e}'_{y_4^b} = -\mathbf{e}_{y_4^b} \mathbf{P}_L \mathbf{E}_{\mathbf{z}_R} \mathbf{E}'_{\mathbf{f}_R}$ and $\mathbf{e}''_{y_4^b} = -\mathbf{e}_{y_4^b} \mathbf{P}_L \mathbf{E}_{\mathbf{z}_R} \mathbf{e}''_{\mathbf{f}_R}$. Now we are ready to prove that the system is dynamic feedback linearizable with respect to $\mathbf{y}^b$. From Sect. 4.6.1, notice that $(\mathbf{y}_1^a, \dot{\mathbf{y}}_1^a, \ddot{\mathbf{y}}_1^a, \mathbf{y}_1^{a(3)})$ is a function of $(\bar{\mathbf{x}}, \mathbf{X}_C^4)$. Furthermore, replacing (4.53) into (4.66) we obtain:

$$\begin{bmatrix} \mathbf{y}_1^{b(4)} \\ \ddot{y}_4^b \\ \ddot{y}_5^b \end{bmatrix} = \mathbf{b}_{\mathbf{y}^b} + \mathbf{E}_{\mathbf{y}^b} \bar{\mathbf{b}} + \mathbf{E}_{\mathbf{y}^b} \bar{\mathbf{E}} \bar{\mathbf{u}} = \bar{\mathbf{b}}_{\mathbf{y}^b}(\bar{\mathbf{x}}, \mathbf{X}_C^2) + \bar{\mathbf{E}}_{\mathbf{y}^b}(\bar{\mathbf{x}}, \mathbf{X}_C^0) \bar{\mathbf{u}}. \tag{4.67}$$

Therefore, we can conclude that $\mathbf{y}^b$ is a dynamic feedback linearizing output for all $\bar{\mathbf{x}}$ such that the decoupling matrix $\bar{\mathbf{E}}_{\mathbf{y}^b}$ is invertible, i.e., such that $\mathbf{E}_{\mathbf{y}^b}$ and $\bar{\mathbf{E}}$ are full rank. From Sect. 4.6.1, $\bar{\mathbf{E}}$ is full rank if $l \neq 0$, $\delta \neq \pm\pi/2$ and $u_1 \neq 0$. On the other hand, $\mathbf{E}_{\mathbf{y}^b}$ is invertible if $\mathbf{e}''_{y_4^b} \neq 0$. Recalling the expression of $\mathbf{e}''_{y_4^b}$, $\mathbf{E}_{\mathbf{y}^b}$ is invertible if

$$(\mathbf{P}_L \mathbf{z}_R)^\top \begin{bmatrix} -\mathbf{e}_3 & \mathbf{0} & \mathbf{e}_1 \end{bmatrix} \mathbf{P}_L (\mathbf{I}_3 - \mathbf{z}_R \mathbf{z}_R^\top) \mathbf{d} \neq 0. \tag{4.68}$$

In order to simplify the notation, let us define $\mathbf{v}_1 = \mathbf{P}_L\mathbf{z}_R$ and $\mathbf{v}_2 = \mathbf{P}_L(\mathbf{I}_3 - \mathbf{z}_R\mathbf{z}_R^\top)\mathbf{d}$. Noticing that $[-\mathbf{e}_3\ \mathbf{0}\ \mathbf{e}_1]$ is the skew symmetric matrix relative to $\mathbf{e}_2$, we have that the previous inequality is equal to $\mathbf{v}_1^\top(\mathbf{e}_2 \times \mathbf{v}_2) \neq 0$, that we can also write as $\mathbf{e}_2^\top(\mathbf{v}_2 \times \mathbf{v}_1) \neq 0$. Since $\mathbf{v}_1$ and $\mathbf{v}_2$ belong to the same plane $\mathcal{P}_L$, $(\mathbf{v}_2 \times \mathbf{v}_1)$ is orthogonal to $\mathcal{P}_L$ and thus parallel to $\mathbf{y}_L$. Notice that $\mathbf{y}_L$ is equal to $\mathbf{e}_2$ in frame $\mathcal{F}_L$. Therefore $\mathbf{e}_2^\top(\mathbf{v}_2 \times \mathbf{v}_1) \neq 0$ if and only if $(\mathbf{v}_2 \times \mathbf{v}_1) \neq \mathbf{0}$.

In order to study the conditions for which $\mathbf{P}_L(\mathbf{I}_3 - \mathbf{z}_R\mathbf{z}_R^\top)\mathbf{d} \times \mathbf{P}_L\mathbf{z}_R \neq 0$, we can show that $\mathbf{P}_L(\mathbf{I}_3 - \mathbf{z}_R\mathbf{z}_R^\top)\mathbf{d} \times \mathbf{P}_L\mathbf{z}_R = \mathbf{P}_L\mathbf{d} \times \mathbf{P}_L\mathbf{z}_R$. Indeed,

$$\mathbf{P}_L(\mathbf{I}_3 - \mathbf{z}_R\mathbf{z}_R^\top)\mathbf{d} \times \mathbf{P}_L\mathbf{z}_R - \mathbf{P}_L\mathbf{d} \times \mathbf{P}_L\mathbf{z}_R = \mathbf{0} \tag{4.69a}$$

$$\left(\mathbf{P}_L(\mathbf{I}_3 - \mathbf{z}_R\mathbf{z}_R^\top)\mathbf{d} - \mathbf{P}_L\mathbf{d}\right) \times \mathbf{P}_L\mathbf{z}_R = \mathbf{0} \tag{4.69b}$$

$$\mathbf{P}_L\mathbf{z}_R\mathbf{z}_R^\top\mathbf{d} \times \mathbf{P}_L\mathbf{z}_R = \mathbf{0}. \tag{4.69c}$$

The previous equivalent conditions always hold since $\mathbf{z}_R\mathbf{z}_R^\top$ is the projector on $\mathbf{z}_R$ and therefore $\mathbf{z}_R\mathbf{z}_R^\top\mathbf{d}$ is always parallel to $\mathbf{P}_L\mathbf{z}_R$ (or zero).

Finally, we have that $\mathbf{E}_{\mathbf{y}^b}$ is invertible if $\mathbf{P}_L\mathbf{d} \times \mathbf{P}_L\mathbf{z}_R \neq \mathbf{0}$, and therefore if $\mathbf{z}_R$ is not perpendicular to the plane $\mathcal{P}_L$ and if $\mathbf{P}_L\mathbf{z}_R$ and $\mathbf{P}_L\mathbf{d}$ are not parallel, i.e., if $\mathbf{z}_R \not\perp \mathcal{P}_L$ and $\mathbf{P}_L\mathbf{z}_R \not\parallel \mathbf{P}_L\mathbf{d}$. Summarizing, the decoupling matrix $\bar{\mathbf{E}}_{\mathbf{y}^b}$ is invertible, and thus $\mathbf{y}^b$ is a feedback linearizing output, for all the states except if $l = 0$ and $\delta = \pm\pi/2$ (singularity of the spherical coordinates), $u_1 = 0$ (singularity of the feedback linearization with respect to $\mathbf{y}^a$), $\mathbf{z}_R \perp \mathcal{P}_L$ and $\mathbf{P}_L\mathbf{z}_R \parallel \mathbf{P}_L\mathbf{d}$.

In the cases in which $\bar{\mathbf{E}}_{\mathbf{y}^b}$ is invertible, using the control law

$$\bar{\mathbf{u}} = \bar{\mathbf{E}}_{\mathbf{y}^b}(\bar{\mathbf{x}}, \mathbf{X}_C^0)^{-1}\left[-\bar{\mathbf{b}}_{\mathbf{y}^b}(\bar{\mathbf{x}}, \mathbf{X}_C^2) + \mathbf{v}^b\right], \tag{4.70}$$

where $\mathbf{v}^b = [v_1^b\ v_2^b\ v_3^b\ v_4^b\ v_5^b]^\top \in \mathbb{R}^5$ are virtual inputs, we obtain

$$y_1^{b(4)} = v_1^b,\quad y_2^{b(4)} = v_2^b,\quad y_3^{b(4)} = v_3^b,\quad y_4^{b(2)} = v_4^b,\quad y_5^{b(2)} = v_5^b. \tag{4.71}$$

Furthermore, the total relative degree with respect to $\mathbf{y}^b$ is $r^b = 16$ that corresponds to the dimension of the extended state $\bar{\mathbf{x}}$ that is $\bar{n} = 16$. Therefore the system is exactly dynamic feedback linearizable and the linearized system (4.71) does not have an internal dynamics.

As done in Sect. 4.6.1, to track a desired trajectory $\mathbf{y}^{bd}(t)$, we can apply to the linearized system (4.71) a linear controller, as the one in (4.56) obtaining a control strategy similar to the one represented in Fig. 4.4.

Proposition *For the analyzed system it exists at least one parametrization η of $\mathbf{R}_R$ and one of its elements η_i such that $\mathbf{y}^b = [l\ \varphi\ \delta\ \vartheta_A\ \eta_i]^\top$ is an exact feedback linearizing output for each state, except if $l = 0$, $\delta = \pm\pi/2$, $u_1 = 0$, $\mathbf{z}_R \perp \mathcal{P}_L$ and $\mathbf{P}_L\mathbf{z}_R \parallel \mathbf{P}_L\mathbf{d}$. Furthermore, considering as input $\bar{\mathbf{u}} = [\ddot{u}_1\ u_2\ u_3\ u_4\ \ddot{u}_5]^\top$, the control law $\bar{\mathbf{u}} = \Gamma_{DFL}^b(\bar{\mathbf{x}}, \mathbf{X}_C^4, \mathbf{y}^{bd}(t))$ defined by (4.70) and (4.56) (properly adapted), exponentially steers $\mathbf{y}^b$ along the desired trajectories $y_i^{bd}(t) \in C^3$ for $i = 1, 2, 3$, and $y_j^{bd}(t) \in C^1$ for $j = 4, 5$.* □

4.6.3 Dynamic Feedback Linearization for Passive Link Actuator

Considering $\mathbf{y}^c$ as output of interest and applying the feedback linearization technique, we need to differentiate each entry of $\mathbf{y}^c$ until the input appears. From (4.16) and (4.44), $\mathbf{y}^c$ has to be differentiated twice to see the input appear:

$$\begin{bmatrix} \ddot{\mathbf{y}}_1^c \\ \ddot{y}_4^c \end{bmatrix} = \begin{bmatrix} \bar{\mathbf{M}}\mathbf{a} \\ b_{\eta_i} \end{bmatrix} + \begin{bmatrix} \bar{\mathbf{M}}\mathbf{Q} \\ \mathbf{e}_{\eta_i} \end{bmatrix} \mathbf{u} = \underbrace{\begin{bmatrix} \bar{\mathbf{M}}\mathbf{a}' \\ b_{\eta_i} \end{bmatrix}}_{\mathbf{b}_{\mathbf{y}^c}(\mathbf{x}, \mathbf{X}_C^2)} + \underbrace{\begin{bmatrix} -\mathbf{J}_\mathbf{q}^\top \mathbf{R}_C^\top \mathbf{R}_R \mathbf{e}_3 & \mathbf{0}_{3\times3} \\ 0 & \mathbf{e}_i^\top \mathbf{T}_\eta \mathbf{J}_R^{-1} \end{bmatrix}}_{\mathbf{E}_{\mathbf{y}^c}(\mathbf{x}, \mathbf{X}_C^0)} \mathbf{u}', \tag{4.72}$$

where $\bar{\mathbf{M}} = \mathbf{M}'^{-1}$ and $\mathbf{a}' = -\mathbf{c}' - \mathbf{g}' - \mathbf{n}' - \mathbf{w}'$. Similarly to the previous cases, the decoupling matrix $\mathbf{E}_{\mathbf{y}^c}$ is singular for every conditions because $\boldsymbol{\tau}_R$ does not appear on $\ddot{\mathbf{y}}_1^c$. This means that the system is not statically feedback linearizable.

As done before, we can apply a dynamic feedback inserting a dynamic compensator in the control u_1. Consider as new input the second derivative of the thrust and the torque, i.e., $\bar{\mathbf{u}}' = [\ddot{u}_1 \;\; \mathbf{u}_2^\top]^\top$. Now $\ddot{\mathbf{y}}_1^c$ has to be differentiated four times to see $\bar{\mathbf{u}}'$ appear, while for y_4^c everything remains the same, indeed:

$$\begin{bmatrix} \ddot{\mathbf{y}}_1^{c(4)} \\ y_4^c \end{bmatrix} = \begin{bmatrix} \ddot{\bar{\mathbf{M}}}(\mathbf{a}' + \mathbf{a}_{\mathbf{u}'}) + 2\dot{\bar{\mathbf{M}}}(\dot{\mathbf{a}}' + \dot{\mathbf{a}}_{\mathbf{u}'}) + \bar{\mathbf{M}}(\ddot{\mathbf{a}}' + \ddot{\mathbf{a}}_{\mathbf{u}'}) \\ b_i + \mathbf{e}_{\eta_i}\bar{\mathbf{u}} \end{bmatrix}, \tag{4.73}$$

where $\ddot{\mathbf{a}}_{\mathbf{u}'}$, after replacing the system dynamics, results:

$$\ddot{\mathbf{a}}_{\mathbf{u}'} = \bar{\ddot{\mathbf{a}}}_{\mathbf{u}'} + \mathbf{J}_\mathbf{q}^\top \mathbf{R}_C^\top \mathbf{R}_R(-\ddot{u}_1 \mathbf{e}_3 - u_1 \left[\mathbf{J}_R^{-1}\boldsymbol{\tau}_R\right]_\times \mathbf{e}_3), \tag{4.74}$$

where $\bar{\ddot{\mathbf{a}}}_{\mathbf{u}'}$ gathers all the terms that do not depend on $\bar{\mathbf{u}}'$. Replacing Eqs. (4.52) and (4.74) into (4.73) we obtain

$$\begin{bmatrix} \mathbf{y}_\mathbf{q}^{(4)} \\ \ddot{y}_4 \end{bmatrix} = \bar{\mathbf{b}}_{\mathbf{y}^c}(\bar{\mathbf{x}}, \mathbf{X}_C^4) + \underbrace{\begin{bmatrix} \bar{\mathbf{M}}\mathbf{J}_\mathbf{q}^\top \mathbf{R}_C^\top \mathbf{R}_R \mathbf{T} & \mathbf{0}_{3\times1} \\ \tilde{\mathbf{e}}_3 & \tilde{e}_2 \end{bmatrix}}_{\bar{\mathbf{E}}_{\mathbf{y}^c}(\bar{\mathbf{x}}, \mathbf{X}_C^0)} \bar{\mathbf{u}}', \tag{4.75}$$

where $\bar{\mathbf{b}}(\bar{\mathbf{x}}, \mathbf{X}_C^4)$ collects all the terms that do not depend on $\bar{\mathbf{u}}'$, $\tilde{e}_2 = \mathbf{e}_i^\top \mathbf{T}_\eta \mathbf{J}_R^{-1} \mathbf{e}_3$, and $\tilde{\mathbf{e}}_3 = \mathbf{e}_i^\top \mathbf{T}_\eta \mathbf{J}_R^{-1} [\mathbf{0}_{3\times1} \;\; \mathbf{e}_1 \;\; \mathbf{e}_2]$. Similar to Sect. 4.6.1, the decoupling matrix $\bar{\mathbf{E}}_{\mathbf{y}^c}$ results to be invertible if $u_1 \neq 0$ and if the parametrization $\boldsymbol{\eta}$ of $\mathbf{R}_R$ and one of its elements η_i are chosen such that $\tilde{e}_2 \neq 0$, i.e., if (4.45) is verified in the domain of interest. Then, in the case in which $\bar{\mathbf{E}}_{\mathbf{y}^c}$ is invertible, defining $\mathbf{v}^c = [v_1^c \;\; v_2^c \;\; v_3^c \;\; v_4^c]^\top \in \mathbb{R}^4$ as virtual inputs, the control law

$$\bar{\mathbf{u}}' = \bar{\mathbf{E}}_{\mathbf{y}^c}(\bar{\mathbf{x}}, \mathbf{X}_C^0)^{-1} \left[-\bar{\mathbf{b}}_{\mathbf{y}^c}(\bar{\mathbf{x}}, \mathbf{X}_C^4) + \mathbf{v}^c\right], \tag{4.76}$$

brings the original system in the equivalent linear one:

$$y_1^{c(4)} = v_1^c, \quad y_2^{c(4)} = v_2^c, \quad y_3^{c(4)} = v_3^c, \quad y_4^{c(2)} = v_4^c. \tag{4.77}$$

This means that the system results to be exactly linearizable through dynamic feedback and the linearized system (4.77) does not have an internal dynamics. Indeed, the total relative degree with respect to $\mathbf{y}^c$ is $r^c = 4 + 4 + 4 + 2 = 14 = \bar{n}$, where $\bar{n}$ is dimension of the extended state $\bar{\mathbf{x}}$.

As done in Sect. 4.6.1, to track a desired trajectory $\mathbf{y}^{cd}(t)$, we can apply to the linearized system (4.77) a linear controller as the one in (4.56) obtaining a control strategy similar to the one represented in Fig. 4.4.

Proposition *For the analyzed system it exists at least one parametrization η of $\mathbf{R}_R$ and one of its elements η_i such that $\mathbf{y}^c = [l \ \varphi \ \delta \ \eta_i]^\top$ is an exact feedback linearizing output for each state, except if $l = 0$, $\delta = \pm\pi/2$, $u_1 = 0$. Furthermore, considering as input $\bar{\mathbf{u}}' = [\ddot{u}_1 \ u_2 \ u_3 \ u_4]^\top$, the control law $\bar{\mathbf{u}}' = \Gamma_{DFL}^c(\bar{\mathbf{x}}, \mathbf{X}_C^4, \mathbf{y}^{cd}(t))$ defined by (4.76) and (4.56) (properly adapted), exponentially steers $\mathbf{y}^c$ along the desired trajectories $y_i^{cd}(t) \in C^3$ for $i = 1, 2, 3$, and $y_4^{cd}(t) \in C^1$.* □

4.6.4 Dynamic Feedback Linearization for the Reduced Model

It is easy to recast all the controllers presented so far for the reduced model presented in Sect. 4.3. Indeed this is a sub-case of the general system. However, we found out that for this reduced morel, the tracking of the output $\mathbf{y}_r^b = [\varphi \ \vartheta_A]^\top$ can be obtained with a simpler static-feedback linearizing controller, $\Gamma_{\mathrm{SFL}r}^b$. For the corresponding details we refer the interested reader to[6] [12]. Notice that the control actions required by this controller may be discontinuous due to, e.g., desired trajectories possessing a discontinuity in the second derivative or simply, in the real case, due to some noise in the measurements. This has to be taken into account in case one would like, e.g., to minimize mechanical vibrations of the link. Furthermore, as recalled in Sect. 3.3.1, one has to keep in mind that discontinuous inputs cannot be performed by the physical system in exam because the acceleration of both propeller rotation and the corresponding air flow cannot be infinite. However, if one needs to enforce smoother inputs it possible to apply a dynamic compensator in order to get a sufficiently smooth control signal.

Table 4.4 summarizes all the controllers designed so far for the general and particular systems.

[6]Notice that in [12] we considered θ as part of the output instead of ϑ_A. However, in the 2D plane, $\vartheta_A = \theta$ so the results are still valid.

Table 4.4 List of designed controllers with the corresponding main features and characteristics

Ctrl.	Cont. outputs	Goal	Method	Type	Singularities
Γ^a_{HC}	$(l, \varphi, \delta, f_L, \eta_i)$	Stabilization	Hierarchical	Static	$f_R = 0$
Γ^b_{HC}	$(l, \varphi, \delta, \vartheta_A, \eta_i)$	Stabilization	Hierarchical	Static	$f_R = 0$, $\mathbf{z}_R \perp \mathcal{P}_L$, $\mathbf{P}_L\mathbf{z}_R \parallel \mathbf{P}_L\mathbf{d}$
Γ^c_{HC}	$(l, \varphi, \delta, \eta_i)$	Stabilization (passive link actuator)	Hierarchical	Static	$f_R = 0$
Γ^a_{DFL}	$(l, \varphi, \delta, f_L, \eta_i)$	Tracking	DFL	Dynamic	$f_R = 0, l = 0$, $\delta = \pm\pi/2$
Γ^b_{DFL}	$(l, \varphi, \delta, \vartheta_A, \eta_i)$	Tracking	DFL	Dynamic	$f_R = 0, l = 0$, $\delta = \pm\pi/2$, $\mathbf{z}_R \perp \mathcal{P}_L$, $\mathbf{P}_L\mathbf{z}_R \parallel \mathbf{P}_L\mathbf{d}$
Γ^c_{DFL}	$(l, \varphi, \delta, \eta_i)$	Tracking (passive link actuator)	DFL	Dynamic	$f_R = 0, l = 0$, $\delta = \pm\pi/2$
$\Gamma^b_{\text{SFL}^r}$	(φ, ϑ_A)	Tracking (reduced model)	SFL	Static	$\vartheta_A = 0$

4.7 State Estimation

In Remark 4.6.1.1 we already highlighted the fact that in order to compute the control action, only the system state and the knowledge of the trajectory of the platform are needed. Assuming that $\mathbf{w}_1 = \mathbf{X}^4_C$ (we recall that $\mathbf{X}^4_C$ is defined in (4.13)) is a priori known or it is estimated/measured on-line by a set of sensors, then only the knowledge of $\mathbf{x}$ is need to close the control loop. One could directly measure $\mathbf{x}$ by using a collection of sensors such as a GPS, cameras, tracking systems etc. In this section we shall demonstrate that exploiting the tautness of the cable (i.e., when $f_L \neq 0$) it is possible to retrieve $\mathbf{x}$ from a standard set of sensors summarized in Table 4.5. In particular, from the moving platform side, we assume to have an encoder that measures the absolute rotation of the link actuator. Assuming no backlash and constant radius r_W, this is equivalent to measure the length of the link, i.e., $w_2 = r_W\vartheta_W = l$. Furthermore, using a gimbal-like system based on two encoders (see [4] for a similar mechanism), we can measure the direction of the link, i.e., $w_3 = \varphi$ and $w_4 = \delta$.

The aerial vehicle is instead equipped with a standard sensory configuration composed by a 3-axis accelerometer, gyroscope and magnetometer mounted on O_R and aligned along the axis of $\mathcal{F}_R$. Recalling the sensor models provided in Sect. 3.4, the accelerometer measures the specific acceleration of O_R in $\mathcal{F}_R$, i.e.:

$$\mathbf{w}_5 = \mathbf{R}_R^\top(\ddot{\mathbf{p}}_R^W + g\mathbf{e}_3). \tag{4.78}$$

The gyroscope measures the angular velocity of $\mathcal{F}_R$ with respect to $\mathcal{F}_W$, expressed in $\mathcal{F}_R$, i.e., $\mathbf{w}_6 = \boldsymbol{\omega}_R$. Finally the magnetometer measures the known unit vector $\mathbf{h}^W \in \mathsf{S}^2$ describing the magnetic field direction expressed in $\mathcal{F}_R$:

$$\mathbf{w}_7 = \mathbf{R}_R^\top \mathbf{h}^W. \tag{4.79}$$

Using this sensory configuration, part of the state is already measured, though it remains to estimate $\mathbf{R}_R$ and $\dot{\mathbf{q}}$. From the accelerometer, replacing (4.5) into (4.78), we obtain

$$f_L \mathbf{R}_R^\top \mathbf{R}_C(\mathbf{w}_1)\mathbf{d}^C(w_3, w_4) = -m_R \mathbf{w}_5 - f_R \mathbf{e}_3. \tag{4.80}$$

We define $w_8 = \|-m_R \mathbf{w}_5 - f_R \mathbf{e}_3\|$. Notice that, since the controller guarantees a taut/compressed link, it must be that $f_L \neq 0$, thus $w_8 = f_L$. Defining $\mathbf{s}_1^R = (-m_R \mathbf{w}_5 - f_R \mathbf{e}_3)/w_8$ and $\mathbf{s}_1^W = \mathbf{R}_C(\mathbf{w}_1)\mathbf{d}^C(w_3, w_4)$ we have that $\mathbf{R}_R \mathbf{s}_1^R = \mathbf{s}_1^W$. Using also the magnetometer we obtain a direct measurement of $\mathbf{R}_R$ in the following way. Under the assumption that $\mathbf{s}_1^R$ and $\mathbf{w}_7$ are not parallel, let us define $\mathbf{s}_2^R = (\mathbf{s}_1^R \times \mathbf{w}_7)/\left\|\mathbf{s}_1^R \times \mathbf{w}_7\right\|$ and $\mathbf{s}_3^R = \mathbf{s}_1^R \times \mathbf{s}_2^R$. We then get

$$\mathbf{R}_R \mathbf{s}_2^R = \mathbf{R}_R(\mathbf{s}_1^R \times \mathbf{w}_7) = \mathbf{s}_1^W \times \mathbf{h}^W = \mathbf{s}_2^W \tag{4.81}$$

$$\mathbf{R}_R \mathbf{s}_3^R = \mathbf{R}_R(\mathbf{s}_1^R \times \mathbf{s}_2^R) = \mathbf{s}_1^W \times \mathbf{s}_2^W = \mathbf{s}_3^W, \tag{4.82}$$

where $\{\mathbf{s}_1^R, \mathbf{s}_2^R, \mathbf{s}_3^R\}$ is an orthonormal basis and $\mathbf{S}^R = [\mathbf{s}_1^R \ \mathbf{s}_2^R \ \mathbf{s}_3^R] \in \mathsf{SO}(3)$. Then, defining $\mathbf{S}^W = [\mathbf{s}_1^W \ \mathbf{s}_2^W \ \mathbf{s}_3^W]$, we obtain a direct measurements of $\mathbf{R}_R$ from the sensors $(\mathbf{w}_1, w_3, w_4, \mathbf{w}_5, \mathbf{w}_7)$ as:

$$\mathbf{R}_R = \mathbf{S}^W \mathbf{S}^{R^\top} = \mathbf{W}_R(\mathbf{w}_1, w_3, w_4, \mathbf{w}_5, \mathbf{w}_7). \tag{4.83}$$

Notice that we can find $\mathbf{R}_R$ only if $\mathbf{d}^W$ and $\mathbf{h}^W$ are not parallel (otherwise $\mathbf{s}_1^W \times \mathbf{h}^W = \mathbf{0}_3$) and if $f_L \neq 0$. Indeed, if $f_L = 0$ the link becomes slack and the aerial vehicle results decoupled from the rest of the system. In this condition is then not

Table 4.5 List of sensors

#	Type	Position	Reference	Measurement
$\mathbf{w}_1$	–	O_C	$\mathcal{F}_W$	$\mathbf{X}_C^4$
w_2	Absolute encoder	O_C	$\mathcal{F}_C$	$\vartheta_W \approx l$
w_3	Absolute encoder	O_C	$\mathcal{F}_C$	φ
w_4	Absolute encoder	O_C	$\mathcal{F}_C$	δ
$\mathbf{w}_5$	Accelerometer	O_R	$\mathcal{F}_R$	$\mathbf{R}_R(\ddot{\mathbf{p}}_R^W - g\mathbf{e}_3)$
$\mathbf{w}_6$	Gyroscope	O_R	$\mathcal{F}_R$	$\boldsymbol{\omega}_R$
$\mathbf{w}_7$	Magnetometer	O_R	$\mathcal{F}_R$	$\mathbf{R}_R \mathbf{h}^W$

possible to estimate the attitude of the vehicle in a direct way. Nevertheless, this is not a practical issue since the proposed controller guarantees any non zero internal force.[7] Regarding the special case in which we ask a zero internal force for just an instant, e.g., passing from tension to compression, we provide a short discussion in Sect. 4.9. Furthermore, the magnetometer can be replaced with any sensor able to measure a known vector in $\mathcal{F}_W$ expressed in $\mathcal{F}_R$ not parallel to $\mathbf{d}^W$. In the presence of noisy measurements one can exploit $\mathbf{W}_R$ and $\mathbf{w}_6$ designing a filter to obtain a better estimation of $\mathbf{R}_R$ and $\boldsymbol{\omega}_R$ [26].

After having shown how to estimate $\mathbf{R}_R$ from the measurements, it remains to estimate $\dot{\mathbf{q}}$. Defining $\mathbf{z} = [l \ \dot{l} \ \varphi \ \dot{\varphi} \ \delta \ \dot{\delta}] \in \mathbb{R}^6$ we can write its dynamics (see (4.11)) and the respective measurements as

$$\begin{aligned} \dot{\mathbf{z}} &= \mathbf{A}\mathbf{z} + \mathbf{B}\boldsymbol{\sigma}(\mathbf{z}, \mathbf{u}_t, \mathbf{R}_R, \mathbf{X}_C^2) \\ \mathbf{w}_\mathbf{z} &= [w_2 \ w_3 \ w_4]^\top = \mathbf{C}\mathbf{z}, \end{aligned} \tag{4.84}$$

where[8] $\mathbf{A} = \text{diag}(\mathbf{A}', \mathbf{A}', \mathbf{A}')$, $\mathbf{B} = \text{diag}(\mathbf{B}', \mathbf{B}', \mathbf{B}')$, $\mathbf{C} = \text{diag}(\mathbf{C}', \mathbf{C}', \mathbf{C}')$, $\mathbf{A}' = \left[\begin{smallmatrix} 0 & 1 \\ 0 & 0 \end{smallmatrix}\right]$, $\mathbf{B}' = \left[\begin{smallmatrix} 0 \\ 1 \end{smallmatrix}\right]$, $\mathbf{C}' = [1 \ 0]$. Thanks to the particular canonical form of (4.84), in order to get an estimation of $\mathbf{z}$, it is possible to apply the following nonlinear *high gain observer* (HGO) (see Sect. 2.4)

$$\dot{\hat{\mathbf{z}}} = \mathbf{A}\hat{\mathbf{z}} + \mathbf{B}\boldsymbol{\sigma}(\hat{\mathbf{z}}, \mathbf{u}_t, \mathbf{W}_R, \mathbf{w}_1) + \mathbf{H}(\mathbf{w}_\mathbf{z} - \mathbf{C}\hat{\mathbf{z}}), \tag{4.85}$$

where $\mathbf{H} = \text{diag}(\mathbf{H}', \mathbf{H}', \mathbf{H}')$ and $\mathbf{H}' = [\frac{\alpha_1}{\epsilon} \ \frac{\alpha_2}{\epsilon^2}]^\top$, with $\epsilon \in \mathbb{R}_{>0}$, and the gains $(\alpha_1, \alpha_2) \in \mathbb{R}_{>0}$ are set such that the roots of $s^2 + \alpha_1 s + \alpha_2$ have negative real part. The gains (α_1, α_2) influence the convergence rate of the estimation.

Summarizing, using the standard sensory configuration of Table 4.5, we were able to achieve the third goal obtaining an estimation of the whole state:

$$\begin{aligned} \hat{l} = \hat{z}_1 \ \hat{\varphi} = \hat{z}_3 \ \hat{\delta} = \hat{z}_5 \ \hat{\mathbf{R}}_R = \mathbf{W}_R \\ \hat{\dot{l}} = \hat{z}_2 \ \hat{\dot{\varphi}} = \hat{z}_4 \ \hat{\dot{\delta}} = \hat{z}_6 \ \hat{\boldsymbol{\omega}}_R = \mathbf{w}_6. \end{aligned} \tag{4.86}$$

4.7.1 Closed Loop Stability

In Sect. 4.6 we saw that the control law Γ^a_{DFL}, Γ^b_{DFL} and Γ^c_{DFL} need only the knowledge of the state and of the trajectory of the platform in order to close the loop. Thus we can use the state estimation provided by the proposed observer as feedback for the controller. Though, since the system is nonlinear, one cannot apply the separation

[7] As already said, in a preliminary phase one can bring the cable in a taut condition using a near hovering control [25]. Then it can be replaced with our controller to maintain the desired tension or compression.

[8] $\text{diag}(\mathbf{X}_1, \ldots, \mathbf{X}_n)$ is a block matrix having on the main block diagonal the matrices $\mathbf{X}_i$, whereas the off-diagonal blocks are zero matrices.

principle like in the linear case. Nevertheless, thanks to the direct measurements of some entry of the state and to the particular kind of translational dynamics, i.e., triangular block dynamics with a direct measurement of the first state of each block, it can be shown that a strong property holds [24]. In fact, since the closed loop system by the state feedback controller is exponentially stable for every state except the its singularities (see Table 4.4), there exist a $\bar{\epsilon}$ such that, for every $0 < \epsilon \leq \bar{\epsilon}$ in (4.85), the closed loop system with the observer is exponentially stable, except for the zero stress case and for the controller singularities [24] (see Sect. 2.4).

4.7.2 Discussion on Platform State Measurement

To obtain a perfect tracking one has to know the derivatives of $\mathbf{p}_C^C(t)$ up to the fourth order and of $\boldsymbol{\omega}_C(t)$ up to the third order (see Sect. 4.6). Although any controller (not only the one proposed here) needs to know (implicitly or explicitly) those variables to obtain a zero tracking error, it is difficult in practice to directly measure the higher-order derivatives.

In order to overcome such issue, some practical techniques could be applied which are here shortly mentioned. If the model and control input of the system are known (e.g., in the case of an autonomous vehicle), an observer can be designed to retrieve the needed derivatives of $\mathbf{p}_C^C$ and $\boldsymbol{\omega}_C$. Without the dynamic model but with a set of measurements of some derivatives, one can use standard tracking technique or, if the trajectory of the vehicle is sufficiently 'low frequency', the missing higher derivatives could be simply assumed negligible and equal to zero. For the last case, in Sect. 5.3 we show that the tracking error remains small and bounded.

4.8 State Estimation for the Reduced Model

As said in the previous section, finding the minimal sensory setup that still allows to retrieve the full state estimation is a very important problem, for safety but also for technical and cost-related problems. Driven by these practical reasons and also by the intrinsic theoretical appeal of solving control problems with minimal sensing, in this section we show that the standard on-board inertial sensor (i.e., an accelerometer plus a gyroscope) is sufficient to estimate the full state of the reduced system presented at the end of Sect. 4.3. We recall that the latter consists of a tethered aerial vehicles constrained to move on a 2D vertical plane, with the link fixed to the ground and characterized by a constant length. Its dynamics is given by (4.18). We then show the design of an exact nonlinear observer for that purpose. Of course the state estimator designed in Sect. 4.7 can be still applied for the 2D case, but the corresponding sensory setup results non minimal. Indeed, the encoder directly measuring the elevation is not needed in this case.

Under the 2D constraints we can redefine the state vector as $\mathbf{x} = [\varphi \ \dot{\varphi} \ \theta \ \dot{\theta}]^\top = [x_1 \ x_2 \ x_3 \ x_4]^\top \in \mathbb{R}^4$ and the input vector as $\mathbf{u} = [f_R \ \tau_{Ry}]^\top = [u_1 \ u_2]^\top$. In view of those definitions we can rewrite the 2D dynamic model (4.18) in a more convenient state space form:

$$\dot{\mathbf{x}} = \begin{bmatrix} x_2 \\ a_1 c_{x_1} \\ x_4 \\ 0 \end{bmatrix} + \begin{bmatrix} 0 & 0 \\ a_2 c_{x_1+x_3} & 0 \\ 0 & 0 \\ 0 & a_3 \end{bmatrix} \mathbf{u}, \tag{4.87}$$

where $a_1 = -g/l$, $a_2 = 1/(m_R l)$, $a_3 = 1/J_{R22}$ are the constant parameters of the dynamical model. As normal in the literature we define $s_\star = \sin(\star)$ and $c_\star = \cos(\star)$. Again, the on-board inertial sensor provides the following measurements:

$$\omega = x_4 \tag{4.88}$$

$$\mathbf{a} = \mathbf{R}_R^\top (\ddot{\mathbf{p}}_R + g\mathbf{z}_W) = \begin{bmatrix} a_x & a_y & a_z \end{bmatrix}^\top. \tag{4.89}$$

From now on we omit the second row which is zero by construction, i.e., we assume $\mathbf{a} = \begin{bmatrix} a_x & a_z \end{bmatrix}^\top$.

Observing the full state of system (4.87) using the partial measurements (4.88), and (4.89) is a nontrivial nonlinear observation problem, where the nonlinearities appear both in the system dynamics and in the measurements. We show in the following how this problem can be successfully tackled. Fist of all exploiting (4.88) we can define the gyroscope measurement as a new input $u_3 = \omega$ that lets us reduce the system dimension from four to three. Then, substituting (4.7) and (4.87) into (4.89) and performing some simple algebraic manipulation on the accelerometer measurement we obtain

$$\begin{bmatrix} \dot{x}_1 \\ \dot{x}_2 \\ \dot{x}_3 \end{bmatrix} = \begin{bmatrix} 0 & 1 & 0 \\ 0 & 0 & 0 \\ 0 & 0 & 0 \end{bmatrix} \begin{bmatrix} x_1 \\ x_2 \\ x_3 \end{bmatrix} + \begin{bmatrix} 0 \\ a_1 c_{x_1} + a_2 c_{x_1+x_3} u_1 \\ u_3 \end{bmatrix} \tag{4.90}$$

$$\mathbf{a} = \begin{bmatrix} l c_{x_1+x_3} \left(x_2^2 + a_1 s_{x_1} + a_2 s_{x_1+x_3} u_1 \right) \\ l s_{x_1+x_3} \left(x_2^2 + a_1 s_{x_1} + a_2 s_{x_1+x_3} u_1 \right) - a_2 u_1 \end{bmatrix}. \tag{4.91}$$

The problem is then 'reduced' to the observation of the state $[x_1 \ x_2 \ x_3]^\top$ from the knowledge of the measurements $\mathbf{a}$ and the inputs $[u_1 \ u_3]^\top$. However this problem is still nonlinear both in the system dynamics and in the measurement map.

Trying to apply the exact nonlinear high gain observer also in this case, the system should be in the canonical form (2.15). Although at first view system (4.90 and 4.91) does not resemble to a system in canonical form, we shall demonstrate in the following that it can be put in that form using a few appropriate nonlinear transformations.

4.8.1 State/Output Transformations and HGO Design

In the following we prove that there exist a change of coordinates from the original state $\mathbf{x}$ to a new state $\mathbf{z} = [z_1\ z_2\ z_3]^\top$ and from the original measurements $\mathbf{a}$ to a new measurement w such that the system (4.90 and 4.91) appears in the canonical form presented in Sect. 2.4. While doing so we highlight also the intuitions that led us to discover this particular change of coordinates.

First, since the term $x_1 + x_3$ occurs frequently in (4.90 and 4.91) a simplifying choice is to assume $z_1 = x_1 + x_3$. With this choice we have $\dot{z}_1 = \dot{x}_1 + \dot{x}_3 = x_2 + u_3$ and therefore it is natural to choose $z_2 = x_2$ to obtain $\dot{z}_1 = z_2 + u_3$, thus matching with the first row of the sought canonical form (2.15).

Now, if we compare the second and third rows of (4.90), with the corresponding rows of the sought canonical form (2.15) we see that: *(i)* in the canonical form (2.15) the state-dependent nonlinearity ϕ appears only in the last row of the dynamics, but, on the other hand *(ii)* in (4.90) the nonlinearity appears already in the second row. Therefore, in order to push this 'undesired' nonlinearity down from the second to the third row we can define $z_3 = \dot{z}_2 = \dot{x}_2 = a_1 c_{x_1} + a_2 c_{x_1+x_3} u_1$. Summarizing, we propose the following change of variables

$$z_1 = x_1 + x_3, \quad z_2 = x_2, \quad z_3 = \dot{x}_2 = a_1 c_{x_1} + a_2 c_{x_1+x_3} u_1, \tag{4.92}$$

that transforms the system (4.90 and 4.91) in the following form

$$\dot{\mathbf{z}} = \underbrace{\begin{bmatrix} 0 & 1 & 0 \\ 0 & 0 & 1 \\ 0 & 0 & 0 \end{bmatrix}}_{\mathbf{A}} \mathbf{z} + \underbrace{\begin{bmatrix} 0 \\ 0 \\ 1 \end{bmatrix}}_{\mathbf{B}} \sigma'(\mathbf{z}, s_{x_1}, u_1, \dot{u}_1, u_3) + \underbrace{\begin{bmatrix} u_3 \\ 0 \\ 0 \end{bmatrix}}_{\boldsymbol{\lambda}(u_3)} \tag{4.93}$$

$$\mathbf{a} = \begin{bmatrix} l c_{z_1} \left(z_2^2 + a_1 s_{x_1} + a_2 s_{z_1} u_1\right) \\ l s_{z_1} \left(z_2^2 + a_1 s_{x_1} + a_2 s_{z_1} u_1\right) - a_2 u_1 \end{bmatrix}, \tag{4.94}$$

where the sole state-dependent nonlinearity $\sigma' = a_1 z_2 s_{x_1} + a_2 c_{z_1} \dot{u}_1 - a_2 s_{z_1}(z_2 + u_3) u_1$ is now appearing in the third row, as desired. Notice that we have, on purpose, left the term s_{x_1} untransformed. In the following we show why this choice is convenient instead of directly computing s_{x_1} from (4.92).

To reach the form of (2.15) it remains to extract a measurement of z_1 from the accelerometer reading. From (4.94), defining

$$\eta = \sqrt{a_x^2 + (a_z + a_2 u_1)^2} = \pm l \left(z_2^2 + a_1 s_{x_1} + a_2 s_{z_1} u_1\right), \tag{4.95}$$

we can obtain a direct measure of z_1 writing

$$w = \operatorname{atan2}\left(\pm a_x/\eta,\ \pm(a_z + a_2 u_1)/\eta\right) = z_1 + k\pi, \tag{4.96}$$

where $k \in \{0, 1\}$. Notice that the transformation is possible only if $\eta \neq 0$. From (4.12), the internal forces expression for the reduced system (4.87) results:

$$f_L = \frac{1}{a_2} x_2^2 + \frac{a_1}{a_2} s_{x_1} + s_{x_1+x_3} u_1. \tag{4.97}$$

Then, from Eqs. (4.97) and (4.95) it results that

$$\eta = \pm \frac{1}{m_R} \left(\frac{1}{a_2} z_2^2 + \frac{a_1}{a_2} s_{x_1} + s_{z_1} u_1 \right) = \pm \frac{f_L}{m_R}, \tag{4.98}$$

thus the transformation (4.96) requires non zero force along the link, $f_L \neq 0$, as in the previous section. Another time, this correspondence highlights, that the condition $f_L \neq 0$ it is not just related to our particular transformation choice but is a structural observability requirement. Indeed, if the force along the link is zero, the aerial vehicle and the link become two independent systems[9] and the onboard inertial sensor is not enough to estimate the entire state. In the design of the observer we need to consider this singularity especially for the cases where the desired link force passes from a tension ($f_L > 0$) to a compression ($f_L < 0$).

The accelerometer is also used to replace s_{x_1} into (4.93) obtaining the sought canonical form(2.15). In particular from (4.95) we can write

$$s_{x_1} = \left(\pm\eta/l - z_2^2 - a_2 s_{z_1} u_1 \right) / a_1. \tag{4.99}$$

With reference to (4.93) we can then write $\sigma' = \sigma(\mathbf{z}, \boldsymbol{\mu})$, i.e., in terms of only $\mathbf{z}$ and known quantities $\boldsymbol{\mu} = [u_1 \ \dot{u}_1 \ u_3 \ \pm\eta]^\top$. Notice that the time derivative of the thrust is needed. This can be computed numerically from u_1 in an approximate way. However, if one of the previous presented dynamic feedback linearizing controllers is used (see Sect. 4.6), $\dot{u}_1$ is an internal state of the controller and so known precisely.

Observe that, the sign to be put in front of η is ambiguous. It is convenient to recast this ambiguity putting always a positive sign and considering two possible values: $\eta_+ = +\eta$ and $\eta_- = -\eta$. Corresponding to these quantities we then get two different dynamic models, the one with $(\sigma(\mathbf{z}, \boldsymbol{\mu}_+), w_+)$ and the one with $(\sigma(\mathbf{z}, \boldsymbol{\mu}_-), w_-)$ corresponding to the use of η_+ and η_-, respectively. At each time instant only one choice for η is correct, i.e., $\eta = f_L/m_R$, to which corresponds the correct measure $w = z_1$ and the correct dynamics. It is not possible, however, to discriminate the correct choice instantaneously from the measurements only. In Sect. 4.8.4 we propose a discriminating solution based on the prediction error.

Although less intuitive, the computation of s_{x_1} into (4.93) exploiting the accelerometer readings, allows to concentrate the ambiguity only on the sign of η obtaining two possible dynamic models. Whereas with the more canonical technique, i.e., inverting the state transformation, we would have four different dynamic models. Indeed

[9] In fact, due to the assumption that the mass and rotational inertia of the link are negligible, the link force represents the only coupling force between the two sub-systems.

from (4.92) we would get $s_{x_1} = \pm\sqrt{1-(z_3 - a_2 c_{z_1} u_1)^2/a_1^2}$ that presents another ambiguity on the sign thus generating four possible combinations of the dynamics and measurement equations.

For the design of the observer let us assume that the correct choice between η_+ and η_- is known. In Sect. 4.8.4 we shall propose a method to gain this knowledge. Under this assumption, the described transformation of the state and the measurements have finally transformed the original system (4.90 and 4.91) in an equivalent one in a canonical form, i.e., $\dot{\mathbf{z}} = \mathbf{A}\mathbf{z} + \mathbf{B}\sigma(\mathbf{z}, \boldsymbol{\mu}) + \boldsymbol{\lambda}(\boldsymbol{\mu})$ and $w = [1\ \ 0\ \ 0]\mathbf{z} = \mathbf{C}\mathbf{z}$, for which we can use the following high gain observer [24]

$$\dot{\hat{\mathbf{z}}} = \mathbf{A}\hat{\mathbf{z}} + \mathbf{B}\sigma(\hat{\mathbf{z}}, \boldsymbol{\mu}) + \boldsymbol{\lambda}(\boldsymbol{\mu}) + \mathbf{H}(w - \mathbf{C}\hat{\mathbf{z}}), \tag{4.100}$$

where $\mathbf{H} = \left[\frac{\alpha_1}{\epsilon}\ \frac{\alpha_2}{\epsilon^2}\ \frac{\alpha_3}{\epsilon^3}\right]^{\mathsf{T}}$, $\epsilon \in \mathbb{R}_{>0}$, and $\alpha_i \in \mathbb{R}_{>0}$ are set such that the roots of $p^3 + \alpha_1 p^2 + \alpha_2 p + \alpha_3$ have negative real part.[10]

4.8.2 Observation of the Original State

From the estimation of $\mathbf{z}$, in order to obtain the estimation of the original state $\mathbf{x}$, we note that the state transformation (4.92) is not directly invertible. One can notice that the only knowledge of $\mathbf{z}$ is not enough to retrieve x_1, indeed from (4.92) one can extract only c_{x_1}. Nevertheless, exploiting also the accelerometer measurements (4.99), we can write

$$\begin{bmatrix} c_{x_1} \\ s_{x_1} \end{bmatrix} = \begin{bmatrix} \frac{1}{a_1}(z_3 - a_2 c_{z_1} u_1) \\ \frac{1}{a_1}\left(\frac{\eta}{l} - z_2^2 - a_2 s_{z_1} u_1\right) \end{bmatrix} = \boldsymbol{h}(\mathbf{z}, u_1, \eta). \tag{4.101}$$

The state x_1 can then be computed as the phase of the unit vector $\boldsymbol{h}(\mathbf{z}, u_1, \eta)$ denoted by $\angle \boldsymbol{h}(\mathbf{z}, u_1, \eta)$. Thus the estimation of the original state is given by

$$\hat{\mathbf{x}} = \hat{\mathbf{x}}(\hat{\mathbf{z}}, u_1, \eta) = \begin{bmatrix} \angle \boldsymbol{h}(\hat{\mathbf{z}}, u_1, \eta) \\ \hat{z}_2 \\ \hat{z}_1 - \angle \boldsymbol{h}(\hat{\mathbf{z}}, u_1, \eta) \\ u_3 \end{bmatrix}. \tag{4.102}$$

The full observer chain is then depicted in Fig. 4.5.

[10] The difference $(w - C\hat{\xi})$ stands here for the unique angle $\beta \in (-\pi, \pi]$ such that $\beta + C\hat{\xi} = w + k2\pi$ for a certain $k \in \mathbb{Z}$.

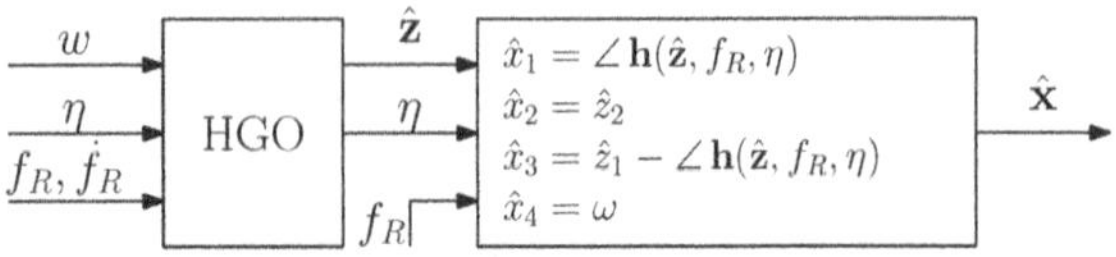

Fig. 4.5 Observer. © 2020 IEEE. Reprinted, with permission, from [12]

4.8.3 Closed-Loop System Stability with State Observation

To prove the stability of the closed loop system when the control action is computed from the estimated state, a similar reasoning to the one in Sect. 4.7 can be done. In particular, for each of the state feedback linearizing controllers, since they are stable out of some singularities (see Table 4.4) there exists $\epsilon^\star > 0$ such that, for every $0 < \epsilon < \epsilon^\star$, the global closed loop system is exponentially convergent for every state except for the zero internal link force and the for the controller singularities. Although the convergence of the observer is almost global, an initialization phase of the estimation can be useful in order to minimize the transient duration, e.g., using the method proposed in [6] for a quasi static initial condition.

Furthermore notice that this observer cannot be directly employed with the static controllers of Table 4.4. Indeed a measure of the derivative of the thrust is needed. Nevertheless one can compute it numerically or applying a dynamic compensator to such controllers. This last option is omitted in this book since it is a trivial extension.

4.8.4 Disambiguation of η and Observability Discussion

As explained before, the transformation of the measurements presents an ambiguity on the sign of η, that can be considered positive, η_+, or negative, η_-. We show in this section how the correct choice can be easily made. Refer to Fig. 4.6 for a graphical representation.

For each of the two possible choices let us implement an HGO equal to (4.100)

$$\dot{\hat{\mathbf{z}}}_+ = \mathbf{A}\hat{\mathbf{z}}_+ + \mathbf{B}\sigma(\hat{\mathbf{z}}_+, \boldsymbol{\mu}_+) + \boldsymbol{\lambda}(\boldsymbol{\mu}_+) + \mathbf{H}(w_+ - \mathbf{C}\hat{\mathbf{z}}_+) \tag{4.103}$$

$$\dot{\hat{\mathbf{z}}}_- = \mathbf{A}\hat{\mathbf{z}}_- + \mathbf{B}\sigma(\hat{\mathbf{z}}_-, \boldsymbol{\mu}_-) + \boldsymbol{\lambda}(\boldsymbol{\mu}_-) + \mathbf{H}(w_- - \mathbf{C}\hat{\mathbf{z}}_-), \tag{4.104}$$

obtaining two different estimations of the state $(\hat{\mathbf{z}}_+, \hat{\mathbf{z}}_-)$, and therefore two different estimations of the original state $(\hat{\mathbf{x}}_+, \hat{\mathbf{x}}_-)$ of which only one is correct. In order to select the correct state we propose a discrimination method based on the comparison of the measurement prediction errors. At the first observer we assign a prediction error $\tilde{e}_+$ smoothed with an exponential discount factor: $\dot{\tilde{e}}_+ = \lambda\left(\left\|\mathbf{a} - \hat{\mathbf{a}}_+\right\| - \tilde{e}_+\right)$, where $\lambda \in \mathbb{R}_{>0}$ sets the discount rate, and $\hat{\mathbf{a}}_+$ is defined as

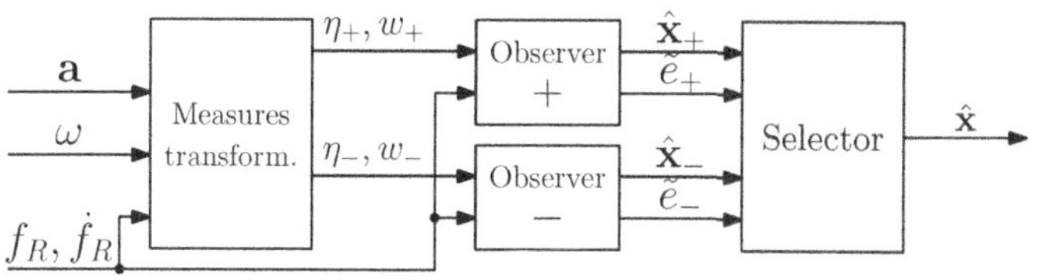

Fig. 4.6 Global observer with disambiguation of η. © 2020 IEEE. Reprinted, with permission, from [12]

$$\hat{\mathbf{a}}_+ = \begin{bmatrix} l c_{\hat{x}_{1+}+\hat{x}_{3+}} \left(\hat{x}_{2+}^2 + a_1 s_{\hat{x}_{1+}} + a_2 s_{\hat{x}_{1+}+\hat{x}_{3+}} u_1\right) \\ l s_{\hat{x}_{1+}+\hat{x}_{3+}} \left(\hat{x}_{2+}^2 + a_1 s_{\hat{x}_{1+}} + a_2 s_{\hat{x}_{1+}+\hat{x}_{3+}} u_1\right) - a_2 u_1 \end{bmatrix}. \tag{4.105}$$

The other prediction error $\tilde{e}_-$ is defined similarly. Then we select the estimation provided by the observer implementation which produces the prediction error closer to zero, i.e., $\hat{\mathbf{x}} = \hat{\mathbf{x}}_+$ if $\tilde{e}_+ \leq \tilde{e}_-$ and $\hat{\mathbf{x}} = \hat{\mathbf{x}}_-$ otherwise.

In Fig. 4.6 we represent the full observer with the discrimination chain. Notice that the disambiguation of the two observers is not done directly using $\hat{\mathbf{z}}$ because is not possible to write the predicted measure $\hat{\mathbf{a}}$ as function of $\hat{\mathbf{z}}$ without introducing another ambiguity. Indeed, as we saw in Sect. 4.8.1, trying to replace $\hat{\mathbf{x}}$ with $\hat{\mathbf{z}}$ into (4.105) inverting the state transformation (4.92), we introduce an ambiguity on the sign of s_{x_1}. Whereas the problem does not hold if we apply this discrimination technique on the original state estimation $(\hat{\mathbf{x}}_+, \hat{\mathbf{x}}_-)$.

Notice that for the motions that implies a constant elevation ($x_2 = 0$), it is not possible to discriminate the correct observer. Indeed, with $x_2 = 0$ and since $w_\pm = z_1 + k\pi$, after a transient the two observers converge to ($\hat{z}_{1\pm} = z_1 + k\pi$, $\hat{z}_{2\pm} = 0$, $\hat{z}_{3\pm} = 0$), and, using (4.102), we obtain ($\hat{x}_{1\pm} = x_1 + k\pi$, $\hat{x}_{2\pm} = 0$, $\hat{x}_{3\pm} = x_3$, $\hat{x}_{4\pm} = x_4$). Under this condition, from Eq. (4.105) it results that the prediction errors of the two observers converge both to zero thus making ineffective the proposed discrimination strategy. Nevertheless, in practice this is not a problem. Indeed, if the controller loop is closed with the wrong observer then the wrong estimation will let the control implement a law that is different from the sole that keeps $x_2 = 0$ causing $x_2 \neq 0$ and thus the predictions errors will become discriminant.

Finally, notice that the ambiguity issue discussed in this section is present only in the initial phases. Whenever the good observer is selected with sufficient certainty, one can switch off the other. For this purpose one can set a confidence threshold on the tracking error of the desired output. If an observer reaches the confidence threshold then this is identified as the correct one and the other one is switched off.

4.9 Discussion on the Proposed Observers

In this book we proposed two observers for the state estimation of a tethered aerial vehicle. One for the general system and one for the reduced one presented in Sect. 4.3.

Table 4.6 List of designed observers with corresponding main features and characteristics

Model	Method	External measurements	Internal measurements	Transformations	Singularities	Degree	Ambiguities
General	HGO	IMU, three encoders	–	Measures	$f_L = 0$	2	No
Reduced	HGO	IMU	$\dot{f}_R$	Measures, state	$f_L = 0$	2	2

The main features and characteristics of such observers are summarized in Table 4.6. Some discussions about the applicability and the robustness of the proposed observers follow.

4.9.1 Applicability

If the link force f_L is zero then some of the measurement transformation shown for both the general and reduced model cannot be determined. In the very special case that the link force has to be constantly equal to zero, the proposed observers are not applicable[11] (while the controller would still work). Nevertheless, the proposed controller Γ^a_{DFL} can guarantee a nonzero link internal force. In the particular case in which the desired link force is passing through zero for a sufficiently short time interval, then the proposed observer can be still used in practice by updating the filter without the correction term in that time instants. For example, for the reduced model we can impose

$$\dot{\hat{\mathbf{z}}} = \mathbf{A}\hat{\mathbf{z}} + \mathbf{B}\sigma(\hat{\mathbf{z}}, \boldsymbol{\mu}) + \boldsymbol{\lambda}(\boldsymbol{\mu}) \quad \text{if } \eta = 0. \tag{4.106}$$

In this way the error dynamics becomes non strictly stable for a short moment but the dynamics returns asymptotically stable as soon as the link force returns to a non-zero value.

4.9.2 Robustness

In order to deal with known drawbacks of the HGO, such as peaking phenomenon and noise sensitivity, many common practical solutions have been presented in the literature, see e.g., [24]. For example, to overcome the peaking phenomenon, it is sufficient to saturate the estimated state on a bounded region that defines the operative regions of the state for the system in exam. In the presence of measurement noise, the use a switched-gain approach can guarantee a quick convergence to the real state during the first phase while reducing the noise effects at steady state.

[11]The static observer in [6] requires a nonzero link force as well to work.

References

1. Oh, S.-R., Pathak, K., Agrawal, S.K., Pota, H.R., Garrett, M.: Approaches for a tether-guided landing of an autonomous helicopter. IEEE Trans. Robot. **22**(3), 536–544 (2006)
2. EC-SAFEMOBIL. EU Coll. Proj. FP7-ICT 288082. http://www.ec-safemobil-project.eu/
3. Sandino, L.A., Bejar, M., Kondak, K., Ollero, A.: Advances in modeling and control of tethered unmanned helicopters to enhance hovering performance. J. Intell. Robot. Syst. **73**(1–4), 3–18 (2014)
4. L. Sandino, D. Santamaria, M. Bejar, A. Viguria, K. Kondak, and A. Ollero. Tether-guided landing of unmanned helicopters without GPS sensors. In: 2014 IEEE International Conference on Robotics and Automation, pp. 3096–3101, Hong Kong, China, May (2014)
5. Sandino, L.A., Bejar, M., Kondak, K., Ollero, A.: A square-root unscented Kalman filter for attitude and relative position estimation of a tethered unmanned helicopter. In: 2015 International Conference on Unmanned Aircraft Systems (ICUAS), pp. 567–576. IEEE (2015)
6. Lupashin, S., D'Andrea, R.: Stabilization of a flying vehicle on a taut tether using inertial sensing. In: 2013 IEEE/RSJ International Conference on Intelligent Robots and Systems, pp. 2432–2438, Tokyo, Japan, Nov (2013)
7. Nicotra, M.M., Naldi, R., Garone, E.: Taut cable control of a tethered UAV. In: 19th IFAC World Congress, pp. 3190–3195, Cape Town, South Africa, Aug (2014)
8. Nicotra, M.M., Naldi, R., Garone, E.: Nonlinear control of a tethered UAV: The taut cable case. Automatica **78**, 174–184 (2017)
9. Siciliano, B., Khatib, O.: Handbook of Robotics. Springer (2008)
10. Tognon, M., Dash, S.S., Franchi, A.: Observer-based control of position and tension for an aerial robot tethered to a moving platform. IEEE Robot. Autom. Lett. **1**(2), 732–737 (2016)
11. Tognon, M., Testa, A., Rossi, E., Franchi, A.: Takeoff and landing on slopes via inclined hovering with a tethered aerial robot. In: 2016 IEEE/RSJ International Conference on Intelligent Robots and Systems, pp. 1702–1707, Daejeon, South Korea, Oct (2016)
12. Tognon, M., Franchi, A.: Dynamics, control, and estimation for aerial robots tethered by cables or bars. IEEE Trans. Robot. **33**(4), 834–845 (2017)
13. Tognon, M., Franchi, A.: Nonlinear observer-based tracking control of link stress and elevation for a tethered aerial robot using inertial-only measurements. In: 2015 IEEE International Conference on Robotics and Automation, pp. 3994–3999, Seattle, WA, May (2015)
14. Tognon, M., Franchi, A.: Position tracking control for an aerial robot passively tethered to an independently moving platform. In: 20th IFAC World Congress, pp. 1069–1074, Toulouse, France, July (2017)
15. Tognon, M., Franchi, A.: Landing and take-off on/from sloped and non-planar surfaces with more than 50 degrees of inclination. In: 9th International Micro Air Vehicles Conference, pp. 97–102 (2017)
16. Tognon, M., Franchi, A.: Control of motion and internal stresses for a chain of two underactuated aerial robots. In: 14th European Control Conference, pp. 1620–1625, Linz, Austria, July (2015)
17. Tognon, M., Franchi, A.: Nonlinear observer for the control of bi-tethered multi aerial robots. In: 2015 IEEE/RSJ International Conference on Intelligent Robots and Systems, pp. 1852–1857, Hamburg, Germany, Sept (2015)
18. Fliess, M., Lévine, J., Martin, P., Rouchon, P.: Flatness and defect of nonlinear systems: introductory theory and examples. Int. J. Control **61**(6), 1327–1361 (1995)
19. Mellinger, D., Kumar, V.: Minimum snap trajectory generation and control for quadrotors. In: 2011 IEEE International Conference on Robotics and Automation, pp. 2520–2525, Shanghai, China, May (2011)
20. Faessler, M., Franchi, A., Scaramuzza, D.: Differential flatness of quadrotor dynamics subject to rotor drag for accurate tracking of high-speed trajectories. IEEE Robot. Autom. Lett. **3**(2), 620–626 (2018)
21. Bullo, F., Lewis, A.D.: Control of Mechanical Systems: Modeling, Analysis, and Design for Simple Mechanical Control Systems, vol. 49. Springer Science Geometric & Business Media (2004)

22. Murray, R.M., Rathinam, M., Sluis, W.: Differential flatness of mechanical control systems: a catalog of prototype systems. In: ASME International Mechanical Engineering Congress and Exposition, San Francisco, CA, Nov (1995)
23. Murray, R.M., Li, Z., Sastry, S.S.: A Mathematical Introduction to Robotic Manipulation. CRC (1994)
24. Khalil, H.K.: Nonlinear Systems, 3rd edn. Prentice Hall (2001)
25. Gioioso, G., Ryll, M., Prattichizzo, D., Bülthoff, H.H., Franchi, A.: Turning a near-hovering controlled quadrotor into a 3D force effector. In: 2014 IEEE International Conference on Robotics and Automation, pp. 6278–6284, Hong Kong, China, May (2014)
26. Mahony, R., Hamel, T., Pflimlin, J.-M.: Nonlinear complementary filters on the special orthogonal group. IEEE Trans. Autom. Control **53**(5), 1203–1218 (2008)

Chapter 5
Simulation and Experimental Results

5.1 Testbed

In the following we shall present the simulation and hardware setups used for the validation of the proposed methods.

5.1.1 Simulation Setup

All the simulations are carried out using Matlab Simulink and considering an aerial vehicle with mass $m_R = 1$[Kg] and inertia $\mathbf{J}_R = 0.25\mathbf{I}_3$[Kg m^2]. We assume constant link actuator radius and inertia equal to $r_W = 0.2$[m] and $J_W = 0.15$[Kg m^2], respectively. The values of the gains and of the desired trajectories are specified in the following, since they are different for each controller.

5.1.2 Hardware Setup

The unidirectional-thrust aerial vehicle used for the experiments consists of a Quadrotor VTOL (see Fig. 5.1), weighting about 1[Kg]. The hardware of the vehicle is the one of a *Mikrokopter*[1] quadrotor. It is endowed with an IMU, and four brushless motor controllers (BLDC ESC) regulating the propeller speed using an in-house developed closed-loop speed controller [1].

We tested our controller with two different setups:

(a) In the first, the quadrotor is linked to a fixed point on the ground by a rigid structure made by carbon-fiber bars and 3D printed parts (see Fig. 5.1a). The

[1]http://www.mikrokopter.de.

M. Tognon and A. Franchi, *Theory and Applications for Control of Aerial Robots in Physical Interaction Through Tethers*, Springer Tracts in Advanced Robotics 140,
https://doi.org/10.1007/978-3-030-48659-4_5

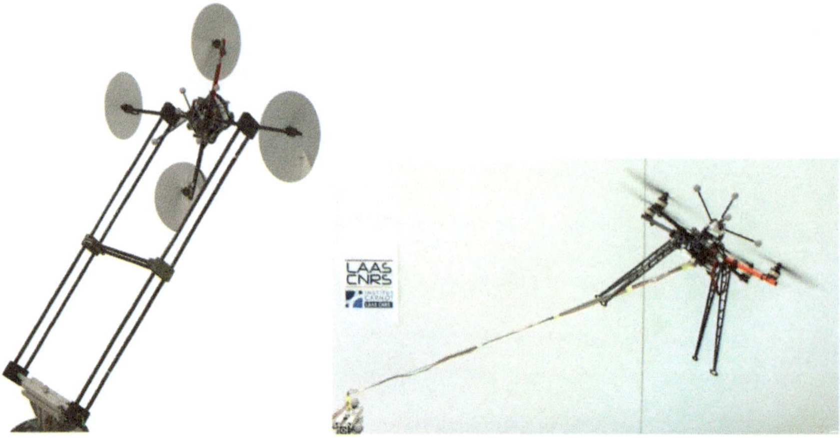

(a) Quadrotor tethered by a bar-link. (b) Quadrotor tethered by a cable-link.

Fig. 5.1 Test bed used for testing the hierarchical controller for a tethered aerial vehicle. On the left the robot is compressing the bar with a force equal to -12 [N], being in an inclined hovering with $\vartheta_A = -80°$. On the right the robot is pulling the cable with a force equal to 7 [N], being in an inclined hovering with $\vartheta_A = 30°$

system implements the reduced model described in Sect. 4.3. In fact, the bar constraints the vehicle to fly on a 2D vertical plane, but, at the same time, does not constrain the vehicle orientation along the $\mathbf{y}_R$ axes. The structure has been designed such that the quadrotor can freely rotate between the two lateral bars without touching them with the propellers. Furthermore the axis of rotation has been brought as close as possible to the vehicle center of mass. Although this setup constraints the vehicle on a reduced space, actually it allows the aerial vehicle to exert on the link both tension and compression.

(b) In the second, the quadrotor is equipped with a light cable with fixed length, ending with a triple hook that can be anchored to a platform (see Fig. 5.1b). The other end of the link is attached to the vehicle as close as possible to its CoM. Once the cable is made taut, the tethered quadrotor can fly on a sphere but can exert only tension on the link.

In both cases the link has a length of 1[m] while a mass of 0.13[Kg] and less than 0.01[Kg] for the first and second setup, respectively, thus negligible w.r.t. the vehicle one.

The control law, implemented in Matlab–Simulink, runs on a desktop PC sending the commanded propeller velocities at 500 [Hz] through a serial communication. The control loop is then closed based on the measurements of: (i) the position and attitude of the vehicle provided at 1 [kHz] by a UKF that fuses the Motion Capture (Mo-Cap) system measurements at 120 [Hz] with the IMU measurements at 1 [kHz]; (ii) the linear and angular velocities of the vehicle provided by the same UKF filter at 1 [KHz].

Table 5.1 Validating experiments

Test bed	Controller	Section	#	Description	Figure
Bar-link (Fig. 5.1a)	Γ^a_{HC}	5.2.1	1	Quasi static conditions, initial tension to vertical compression	5.2
			2	Dynamic trajectory with vertical compression	5.3
Cable-link (Fig. 5.1b)	Γ^b_{HC}	5.2.2	3	Time varying trajectory on φ	5.4
			4	Time varying trajectory on ϑ_A	5.5
			5	Time varying trajectory on both φ and ϑ_A	5.6

5.2 Hierarchical Controllers: Experimental Validation

In order to validate and test the performance of the proposed hierarchical controllers we tried to track some dynamical trajectories showing the ability to independently track the entries of $\mathbf{y}^a$ and $\mathbf{y}^b$. The validation has been done by real experiments with the platform described in Sect. 5.1.2, where the other end of the link has been anchored to a very heavy load (much more than the total lifting of the vehicle). Figure 5.1 shows the corresponding test beds.

Table 5.1 gathers the executed experiments specifying the corresponding test bed, controller and giving a short description. The reader can choose to go directly to the section of a specific experiment or to the corresponding plots.

5.2.1 *Hierarchical Controller for* $\mathbf{y}^a$

For these first tests we use the setup of Fig. 5.1a in order to show the ability of the robot to apply forces to the link, and so to the ground, that go from tension to compression and vice-versa, while changing its position. We shall show such capability in two conditions:

(1) The first is a quasi static condition. The robot is asked to follow a smooth trajectory $\mathbf{y}^{ad}(t) = [\varphi^d(t)\ \ 0\ \ 0\ \ f_L{}^d(t)]^\top$ with $t \in [0, T]$ where T is the final time and $\varphi^d(0) = 20\ [°]$, $\varphi^d(T) = 90°$, $f_L{}^d(0) = 5$ [N], $f_L{}^d(T) = -20N$. Figure 5.2 shows the corresponding results and performed motion. One can notice how the robot is able to keep the bar vertical while pushing it. Since the desired compression is grater then the gravitational one, the robot has to turn and push the bar with an upside down orientation. Even in this unusual configuration for a standard qudrotor, the controlled system remains stable.

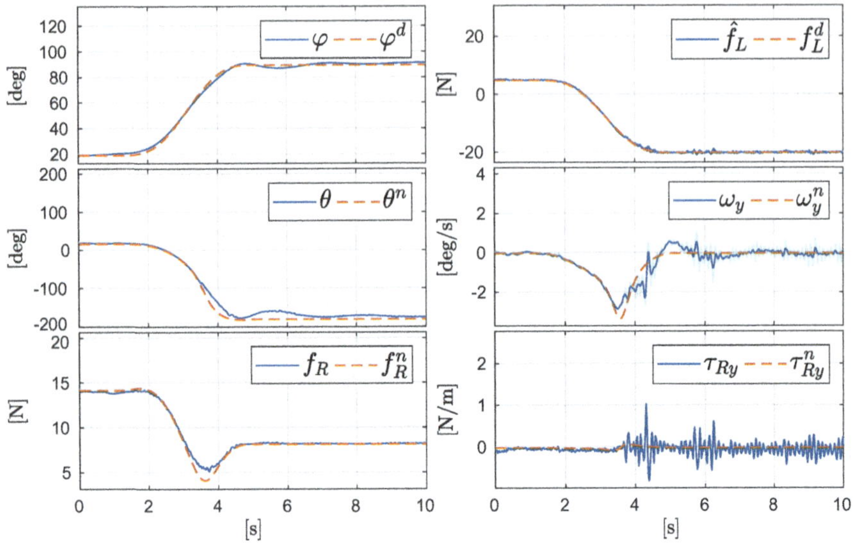

(a) Outputs, state and inputs evolution. Only the variables along the axes of motions are shown.

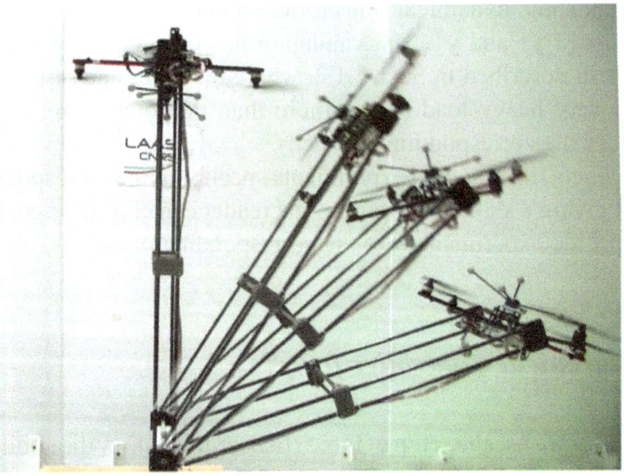

(b) Sequence of snapshots of the experiment.

Fig. 5.2 Results of the the experiment 1 with the hierarchical controller for $\mathbf{y}^a$

(2) The second is a more dynamic trajectory. Like before, $\mathbf{y}^{ad}(t) = [\varphi^d(t)\ 0\ 0\ {f_L}^d(t)]^\top$ is such that $\varphi^d(0) = 40$ [°], $\varphi^d(T/2) = 90°$, $\varphi^d(T/2) = 140°$, and ${f_L}^d(0) = 5$ [N], ${f_L}^d(T/2) = -20N$, ${f_L}^d(T) = 5$ [N]. Figure 5.3 shows the corresponding results and performed motion. In Fig. 5.3b one can notice that to follow the desired trajectory the robot has to flip performing a turn of more than 380 [°] along $\mathbf{y}_R$. We remark that this is not something pre-planned but it is a by-product of the desired internal force trajectory. Although the trajectory is really

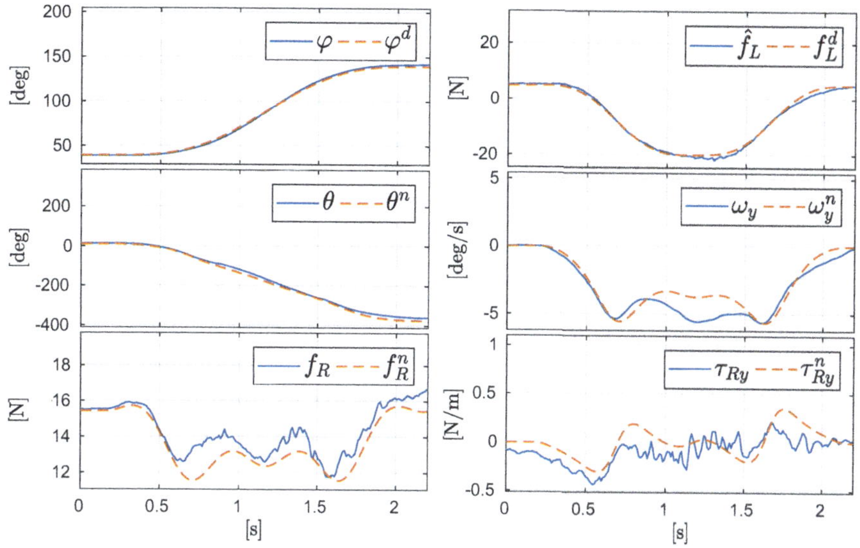

(a) Outputs, state and inputs evolution. Only the variables along the axes of motions are shown.

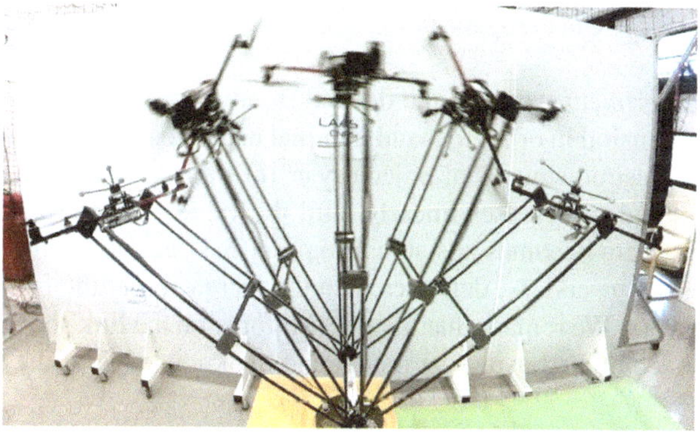

(b) Sequence of snapshots of the experiment.

Fig. 5.3 Results of the the experiment 2 with the hierarchical controller for $\mathbf{y}^a$

acrobatic, the controlled vehicle is able to track the trajectory with sufficient precision.

5.2.2 *Hierarchical Controller for* $\mathbf{y}^b$

Since our control methods works only for the tethered system (nonzero internal force), a pre-tensioning phase is needed. During this phase, the robot is controlled with a standard position controller trying to reach a position outside of the feasible sphere. As soon as, at time t_0 the link is taut (detectable using a threshold in the position error) the controller is activated.

In the following we shall show the results of the control action for three different sinusoidal trajectories:

(1) sinusoidal trajectory with time varying frequency on φ while keeping ϑ_A constant,
(2) sinusoidal trajectory with time varying frequency on ϑ_A while keeping φ constant, and
(3) sinusoidal trajectory with fixed frequency on both φ and ϑ_A,

while δ and ψ are kept constantly to zero. We recall that since there is not a link actuator the length of the link cannot be controlled. The first two tests are done firstly to see that the proposed controller can track a desired trajectory on φ or ϑ_A, independently. Secondly we want to show which is the maximum feasible frequency for both dynamics. Notice that with the validation of controller Γ^b_{HC} for the tracking of $\mathbf{y}^b$, we indirectly validate the controller Γ^a_{HC} for the tracking of $\mathbf{y}^a$, as well. Indeed, given the structure of Γ^b_{HC}, it internally uses Γ^a_{HC} (see (4.43)). We recall that the desired trajectory $\mathbf{y}^{bd}(t)$ is transformed into a new desired trajectory $\mathbf{y}^{ad}(t) = \mathbf{g}_a(\mathbf{y}^{bd}(t), \mathbf{X}^2_C)$ that is tracked by Γ^a_{HC}. Checking the tracking errors of both $\mathbf{y}^b$ and $\mathbf{y}^a$, we can evaluate both controllers.

In the *first experiment* we fixed the desired ϑ^d_A at 15[°]. In this way we assure a sufficiently high tension in order to avoid nominal negative tension values during the experiment. The desired sinusoidal trajectory $\varphi^d(t)$ starts with a frequency equal to $\omega_\varphi = \frac{2\pi}{4}$ [rad/s] and it increases linearly until the value of about $\omega_\varphi = \frac{4\pi}{5}$ [rad/s] after which the system becomes unstable. From Fig. 5.4 one can see that the tracking of φ and ϑ_A, thus of φ and f_L, degrades with the increasing of the frequency of the sinusoidal trajectory. We remark that the internal force on the link shown in the plots is an estimation, computed using the model and the knowledge of the state and the input.

The *second experiment* is the dual, indeed we propose a sinusoidal desired trajectory with varying frequency on ϑ_A while keeping a desired constant $\varphi^d = 45$[°]. For what concerns the frequency of the sinusoidal desired trajectory $\vartheta^d_A(t)$, it starts from a value of $\omega_{\vartheta_A} = \frac{2\pi}{6}$ [rad/s] and increase up to a value of about $\omega_{\vartheta_A} = \frac{8\pi}{9}$ [rad/s]. After that, as it is possible to see from the plots in Fig. 5.5, the tracking error becomes very high. However, the system remains always stable.

Finally, for the *third experiment*, we gave as reference a sinusoidal trajectory on both φ and ϑ_A. The two signals have different frequency and phases, in particular $\omega_\varphi = \frac{2\pi}{4}$ [rad/s] and $\omega_{\vartheta_A} = \frac{2\pi}{6}$ [rad/s], respectively. The results can be seen in Fig. 5.6. As one can see, the trajectories are both tracked with a sufficiently small error. This analysis finally shows that the proposed controller is able to independently track sufficiently slow time varying desired trajectories of $\mathbf{y}^b$ with small tracking errors. On the other hand, as expected, the controller shows increasing tracking errors when asked to follow more dynamic trajectories, revealing its limitations. Anyway, a time varying reference governor (see [2] and references therein) could be applied

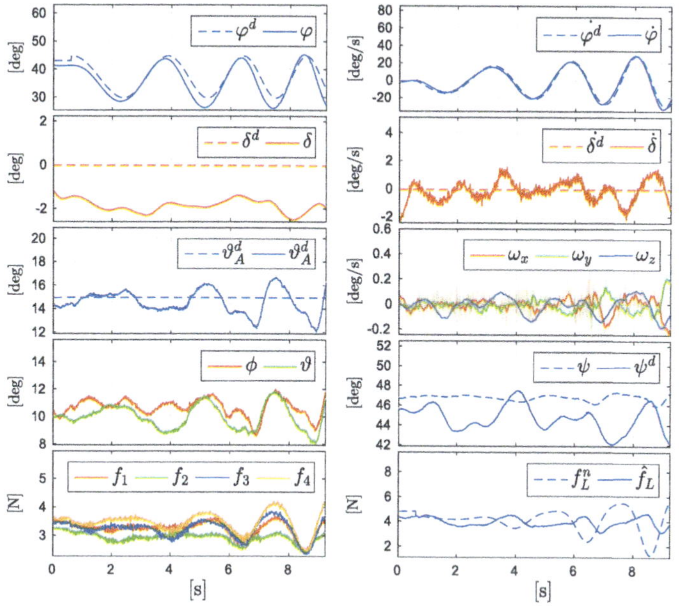

Fig. 5.4 Experimental results: tracking of a sinusoidal input on elevation with varying period with fixed attitude. $f_L{}^n$ is the nominal internal force computed by the flatness from ϑ_A^d. $(f_1, \ldots, f_4)$ are the forces produced by each thruster

to improve tracking performance. We did not report the results of the tracking of δ because they are analogous to the ones related to φ. We also encourage the reader to watch the first part of [3] where some static inclined hovering for a tethered aerial vehicle are shown.

5.3 DFL-Controller for $\mathbf{y}^a$ with Observer

In this section we consider the generic system described in Sect. 4.3 with a cable-like link (only positive internal forces are allowed.). Like in a real patrolling task, the platform follows a certain trajectory in the 3D space mimicking, e.g., a ground robot following a road. We require the aerial vehicle at time t_0 to takeoff from the moving platform, at time t_{circ} to circle above the platform at a certain altitude, and at time t_{land} to land on the moving platform. The yaw angle of the aerial vehicle has to follow the one of the platform. Notice that takeoff and landing are performed while the platform is moving, making these standard maneuvers non trivial.

We firstly test the controller Γ^a_{DFL} for the tracking of $\mathbf{y}^a$, together with the observer designed for the generic system (see Table 4.6). We set $\mathbf{k}_i$ and $\mathbf{k}_j$ such that the error dynamics $\boldsymbol{\xi}_i$ and $\boldsymbol{\xi}_j$ have poles in $(-1, -2, -3, -4)$ and $(-1, -2)$ respectively.

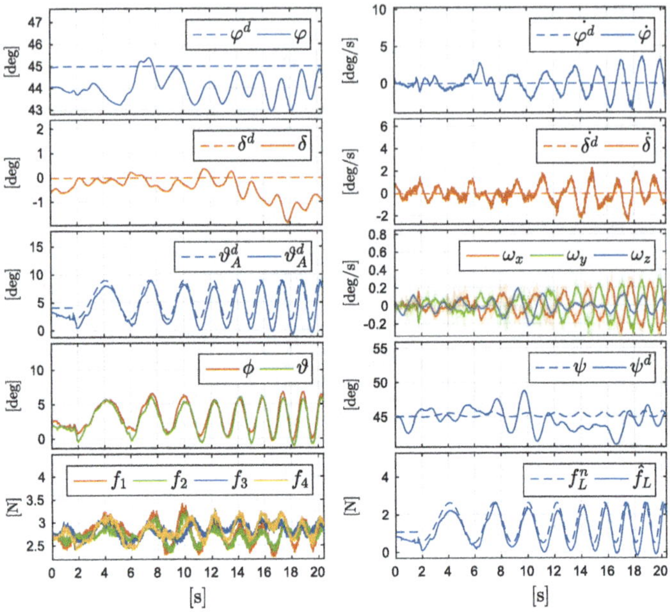

Fig. 5.5 Tracking of a desired sinusoidal trajectory of ϑ_A with varying frequency and fixed φ

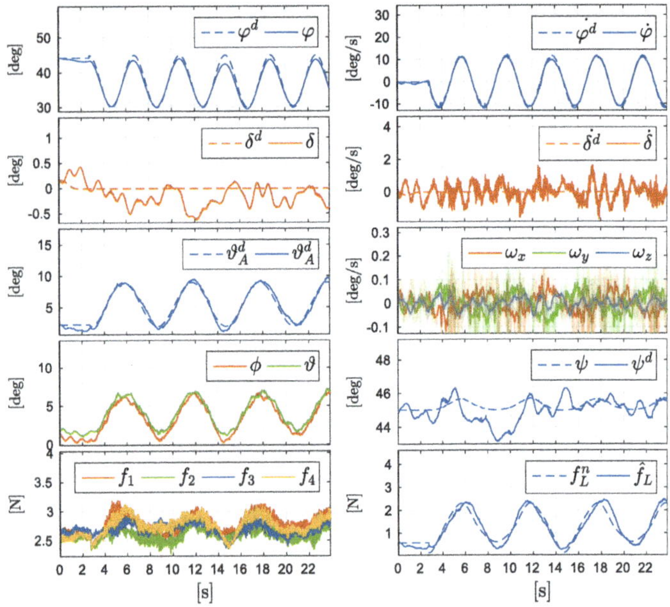

Fig. 5.6 Experimental results: tracking of a desired sinusoidal trajectory on both φ and ϑ_A with fixed period

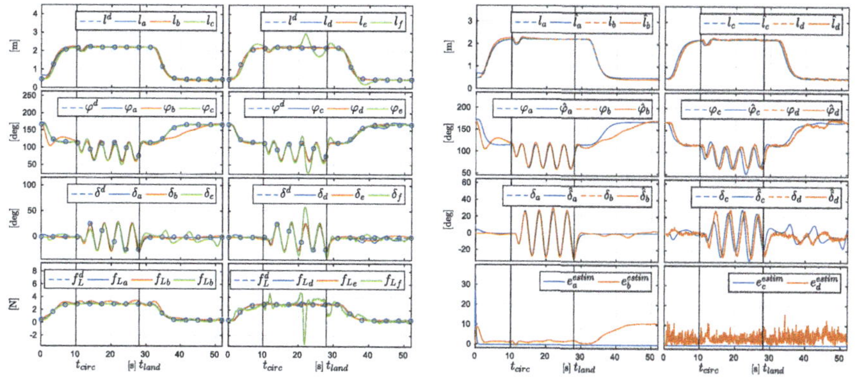

(a) Controller performance: the tracking of the output of interest for each case is plotted. The subscript of the variables indicates the corresponding case. The blue dots highlight the desired trajectory.

(b) Observer performance. The *cases* e) is not plotted for space limitation. However it can be found in Sec. 5.3.5.

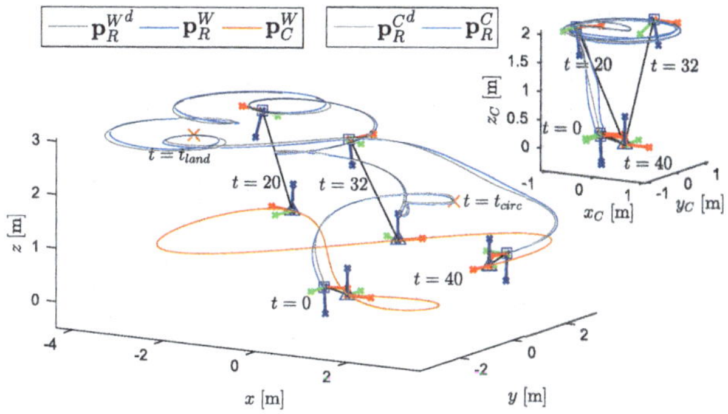

(c) Trajectories visualization for *Case d)*. Left: trajectories in the world frame. Right: aerial vehicle trajectory in the moving platform frame. Red line: platform trajectory. Blue line: aerial robot trajectory. Gray line: desired aerial robot trajectory.

Fig. 5.7 Simulation results: plausible task trajectory. The performance for each non-ideal case are compared.

For the observer we choose $\epsilon = 0.1$ and (α_1, α_2) such that $s^2 + \alpha_1 s + \alpha_2$ has roots $(-3, -4)$. Those values guarantee the stability and ensure a sufficiently fast exponential tracking. During the takeoff the desired tension must go from a small initial tension of 0.5[N] to a steady-state value of 3[N], that is kept for the whole of the circling phase, and then has to go back to the initial value during the landing.

To fully validate our method for real applications we test the convergence and the robustness for different *non ideal* cases commented in the following. Figure 5.7 gathers the main results.

(a) With an *initial position and estimation errors*, after the convergence of the observer (less than one second) the outputs follow the desired trajectories with high fidelity. An animation of this simulation is available at [5].
(b) With a *parametric variation of* 5% we notice a small constant error in the estimation of the state, but we obtain a monotonically decreasing tracking error thanks to the addition of an integral term in the outer loop (4.56), e.g., $v_1 = y_1^{d(4)} + \mathbf{k}_1^{\top}\xi_1 + k_{I1}\int_0^{\top}\xi_1(\tau)d\tau$.
(c) For a moving platform a standard sensory set (e.g., optical flow, IMU and magnetometer) usually is sufficient to measure its trajectory variables up to $\ddot{\mathbf{p}}_C^C$ and ω_C. In this case, in which we have a *partial knowledge of the platform motion*, we can consider as zero the higher derivatives. We observe that the estimation and tracking errors are very small and remain always bounded under a reasonable threshold.
(d) In the presence of *Gaussian noise in the measurements* with typical variance values, we notice that the state estimate becomes slightly noisy but the error remains bounded within small values. The non zero estimation error implies a non zero but bounded tracking error as well.
(e) Since in practice one cannot assume the link attached exactly to O_R, we tested the method for a *vertical offset* of 5[cm] with respect to $\mathcal{F}_R$. In this case the tracking error does not go to zero but remains bounded below a small threshold.
(f) We also compared the dynamic feedback linearizing controller with the hierarchical one. We noticed that to obtain good tracking performance, the hierarchical controller requires very high gains that cause instability in the presence of the same noisy measurements of case (d). Therefore we lowered the gains until we obtained a stable behavior. However these gains are not enough to obtain good tracking performance anymore. Moreover notice that the cable becomes even slack ($f_{L_f} < 0$). Further discussions about the hierarchical control will follow in the next section.

In the following section we provide additional plots and discussions for the previous non ideal cases. We also consider other additional non idealities such as non diagonal inertia matrix, saturation of the inputs and non ideal motors.

For each case we show the control performances plotting the tracking of each output of interest, the global tracking error ξ_{track} computed as the sum of each errors, and the inputs. Concerning f_R and τ_W we also show the nominal input coming from the flatness, $f_R{}^n$ and τ_W^n, that should be applied to obtain the desired output tracking in the nominal case. We also show the observer performances comparing the estimated state and the actual one. The estimation error $e^{estimation}$ is simply calculated as the sum of the estimation error for each entry of the state. Finally we display the trajectories of the aerial vehicle and of the moving platform in the world frame and with respect to $\mathcal{F}_C$. In the 3D plots the position of the moving platform and of the aerial vehicle in some particular instants are represented with a triangle and a square respectively.

5.3.1 Initial Errors

In this section we want to show the closed loop stability of the system in dynamic condition even with some initialization error. The system starts with an error on l of 0.1[m], on φ and δ of 2[°] and on f_L of 0.5[N]. Similarly the initialization of the observer is done with an error of 0.2[m] on $\hat{l}$, of 5[°] on $\hat{\varphi}$ and $\hat{\delta}$, while their velocity are initialized to zero.

In Fig. 5.8 one can see that after the convergence of the observer, that takes less than one second, the controller exponentially steers the outputs along the desired trajectories, while the moving platform is following its own dynamic trajectory.

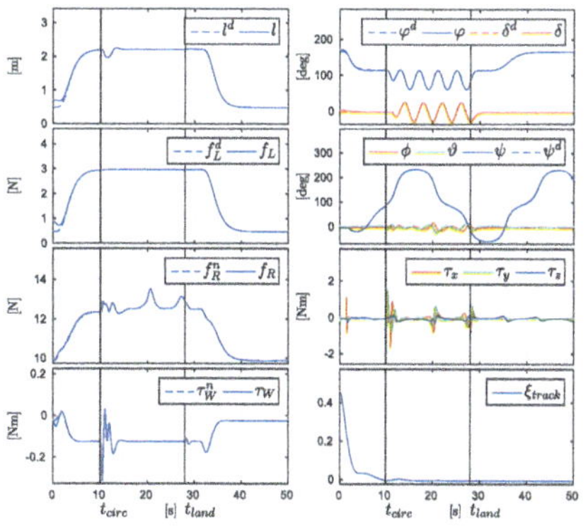

(a) Controller performances.

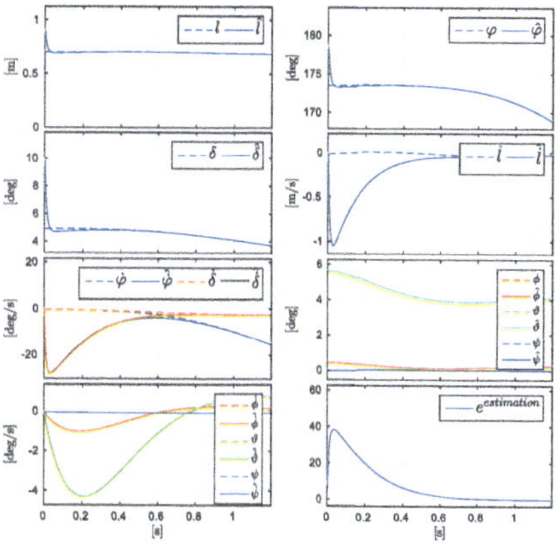

(b) Observer performances. We show here only the first second because after the estimated state follows the actual one with high fidelity for all the remaining simulation.

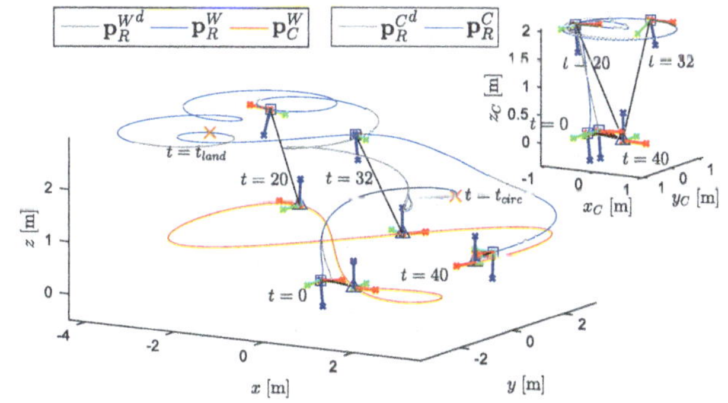

(c) Trajectories visualization.

Fig. 5.8 Simulation: initial errors

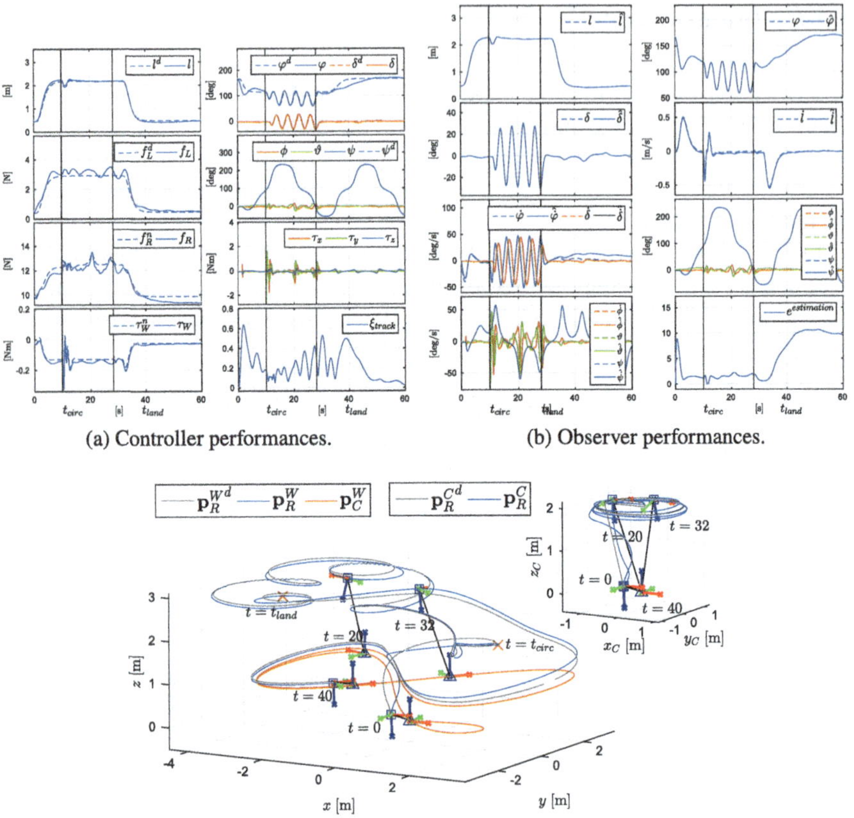

(a) Controller performances.

(b) Observer performances.

(c) Trajectories visualization.

Fig. 5.9 Simulation: parametric variations

5.3.2 *Parametric Variations*

The purpose of the next sections is to investigate the robustness of the proposed method. In particular in this one, we consider some parameter variation between the real model and controller/observer. Indeed in a real scenario we can not know exactly each parameter of the system, thus the controller and observer would be based on some parameter value different from the real one.

Figure 5.9 displays the results of the simulation with a parametric variation of the 5% for each entry, i.e., m_R, $\mathbf{J}_R$, J_W and r_W. In order to partially compensate the effects of the uncertainties we added in the controller an integral term with gain $k_I = 3$.

We can notice that due to the uncertainty of the model we have some nonzero errors in the tracking and in the estimation of the state. Nevertheless the error system remains stable and thanks to the integrator terms, during the landing maneuver we obtained a decreasing tracking error that allows a correct landing of the aerial vehicle.

We performed additional extensive simulations in which we observed that the system remains stable up to a parametric variation of the 20%, after this value the system results unstable. However notice that in reality those parameters are very well measurable with small errors, certainly lower than the 20%.

5.3.3 Limited Knowledge of $\mathbf{p}_C^W(t)$

In Sect. 4.6.1 we saw that the knowledge of $\mathbf{X}_C^4$ is needed in order to compute the control action. In other words, to obtain a perfect tracking one has to know the derivative of $\mathbf{p}_C^C(t)$ up to the fourth order and of $\boldsymbol{\omega}_C(t)$ up to the third order. Although those variables have to be known to obtain zero tracking error, actually, without a posteriori knowledge of the trajectory or the model and control inputs of the system, it is difficult to measure the higher-order derivatives. Nevertheless, in this section we want to show that even with only a partial measurement of $\mathbf{x}_C^4$ the system stays stable and the tracking error remains bounded.

Indeed, for a real moving platform, a standard onboard sensorial configuration, such as optical flow, IMU and magnetometer, is sufficient to obtain $\boldsymbol{\omega}_C(t)$ and $\mathbf{p}_C^C(t)$ up to its second derivative.

In Fig. 5.10b we can observe that the estimation error is almost constantly zero even if $\dot{\boldsymbol{\omega}}_C$ is assumed zero. While in Fig. 5.10a one can notice that the outputs oscillates around the desired value and the tracking error does not go to zero but remains bounded under a reasonable threshold. Nevertheless, with a more "aggressive" platform trajectory the negative effects would be more significants. In Fig. 5.11 the entries of $\mathbf{x}_C^i$ for $i = 1, 2, 3, 4$ are plotted. The last five entries are assumed zero by the observer and the controller.

5.3.4 Noise on the Measurements

In this section we investigate the robustness of the proposed method with the presence of noise in the measurements. Table 5.2 gathers the variance magnitude set for each measurement. For the encoder and the gyroscope we set some reasonable value found in the literature [6]. On the other hand, instead of adding noise on $\mathbf{w}_5$ and $\mathbf{w}_7$ we preferred inserting the noise directly in the measure of the rotational matrix $\mathbf{R}_R$, i.e., in $\mathbf{W}_R$. This is done because the direct measure of $\mathbf{R}_R$ using the accelerometer and the magnetometer is normally filtered with the gyroscope [7], in order to obtain a less noisy estimation of both $\mathbf{R}_R$ and $\boldsymbol{\omega}_R$. The noise added directly to $\mathbf{R}_R$ is comparable to the one we would obtain after the filtering.

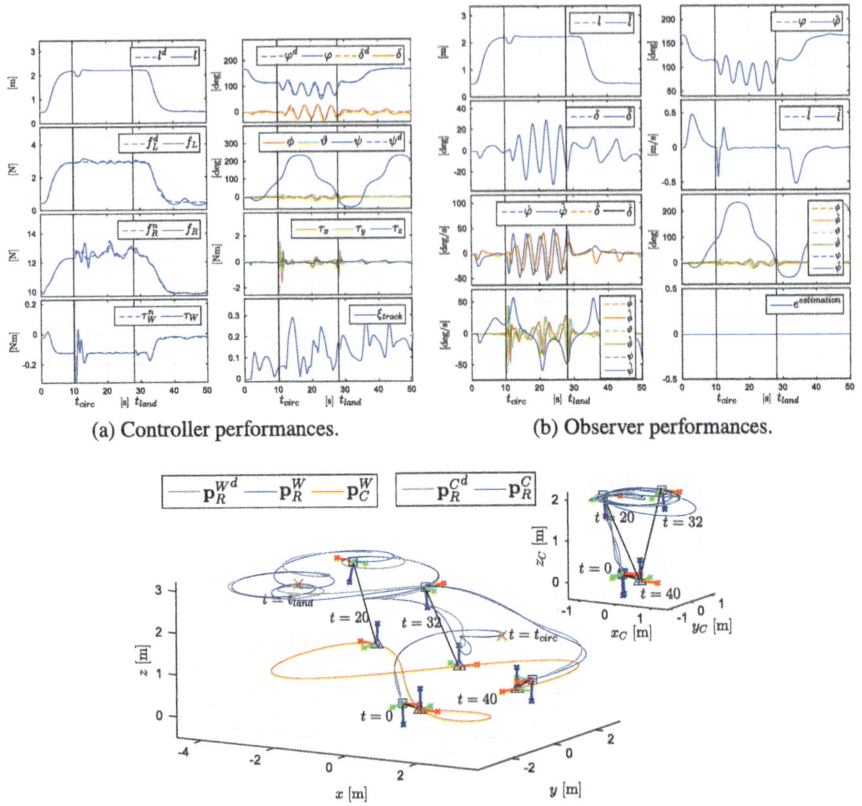

(a) Controller performances.

(b) Observer performances.

(c) Trajectories visualization.

Fig. 5.10 Simulation: limited measurements of the moving platform trajectory

From Fig. 5.12 we can observe that the estimated state shows some noise but the corresponding error remains limited. Due to the noisy component on the estimated state the outputs presents some oscillation as well, especially on the stress that seams to be the more sensitive output to the noise. Nevertheless the tracking error remains small and always bounded.

5.3.5 *Tethered Offset*

Exact attachment of the link to the center of mass of the aerial vehicle is practically unfeasible. Therefore there will always be a non zero offset, although small, between the tether attachment and the center of gravity. This offset makes the translational and rotational dynamics of the aerial robot coupled and can potentially lead to the instability of the controlled system. In this section we want to show the robustness of

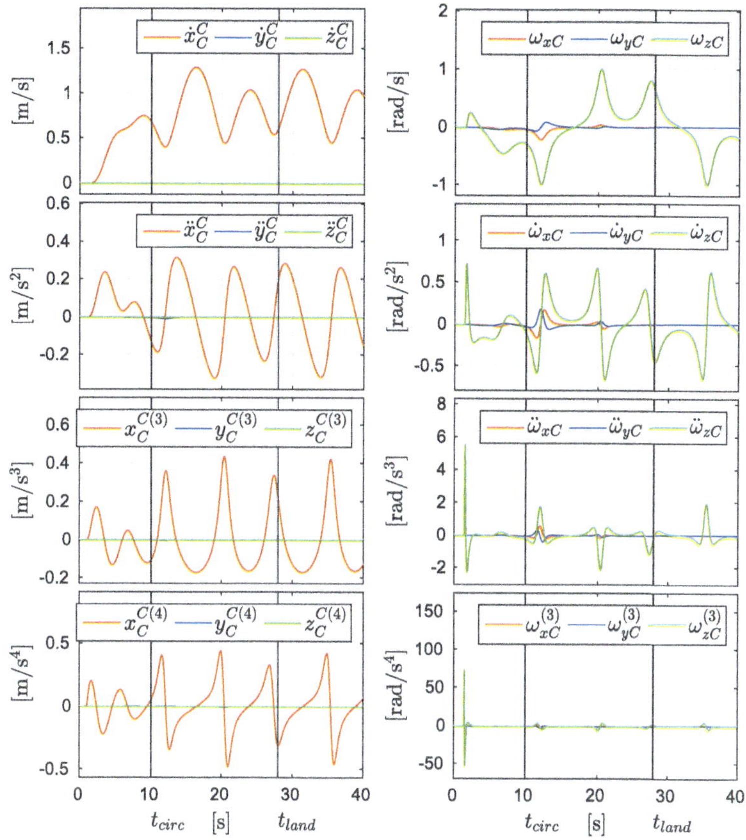

Fig. 5.11 Plot of $\mathbf{x}_C^i$, for $i = 1, 2, 3, 4$. In Simulation (c) all the variables in the last five plots are considered zero by the controller and the observer

Table 5.2 List of sensors

#	Type	Measurement	Noise variance
w_2	Abs. encoder	$\vartheta_W \approx l$	0.008[rad]
w_3	Abs. encoder	φ	0.008[rad]
w_4	Abs. encoder	δ	0.008[rad]
$\mathbf{w}_5$	Accelerometer	$\mathbf{R}_R(\ddot{\mathbf{p}}_R^W + g\mathbf{e}_3)$	–
$\mathbf{w}_6$	Gyroscope	$\boldsymbol{\omega}_R$	0.01[rad/s]
$\mathbf{w}_7$	Magnetometer	$\mathbf{R}_R\mathbf{h}^W$	–
$\mathbf{W}_R$	Complementary filter	$\mathbf{R}_R$	0.001

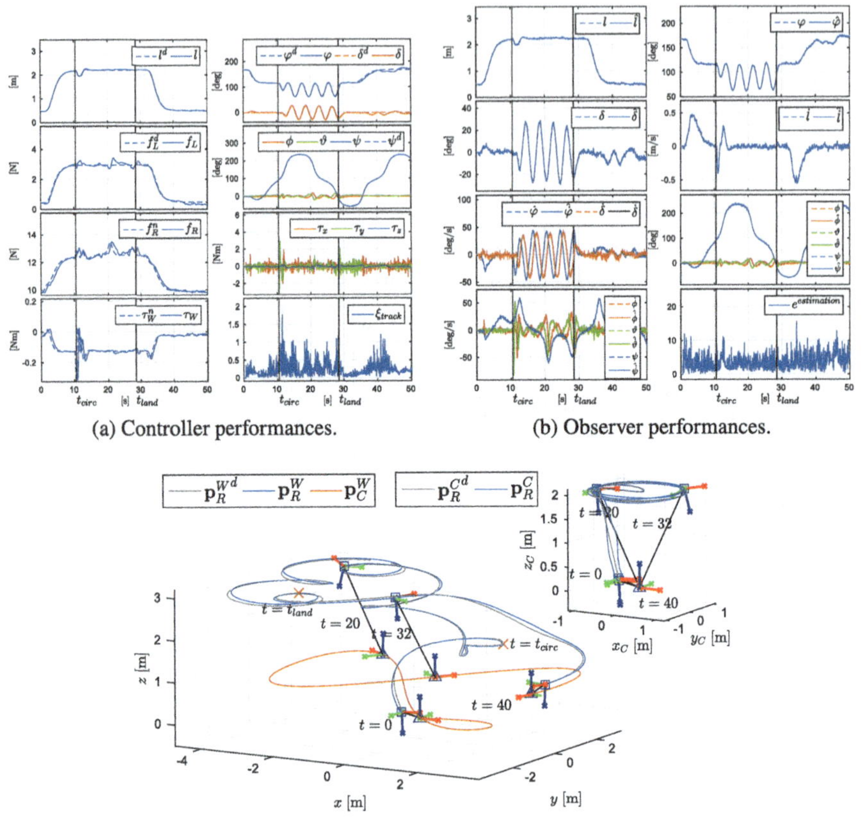

(a) Controller performances.

(b) Observer performances.

(c) Trajectories visualization.

Fig. 5.12 Simulation: noisy measurements

the proposed method when the distance between the attaching point and the center of gravity of the aerial vehicle is non zero (Fig. 5.13). In particular in this simulation the link is attached 5 [cm] vertically below O_R with respect to $\mathcal{F}_R$. As expected, the tracking error does not go to zero but however remains bounded, showing good tracking performances. Notice that the error is higher during the circling phase since this part of the global trajectory is very dynamical and the unmodeled effects due to the offset are larger. However we remark that a good mechanical design could make the tracking error almost negligible.

We tested the method with even larger offsets and we saw that the system remains stable up to a vertical offset of 30[cm], that is an exaggerated value for the system considered in the simulation (small-size quadrotor like vehicle). In fact, note that a larger quadrotor means a larger inertia which actually reduces the negative effects of the offset. In additional simulations, which are not reported here for the sake of brevity, we also tested the robustness of the method with a more general offset

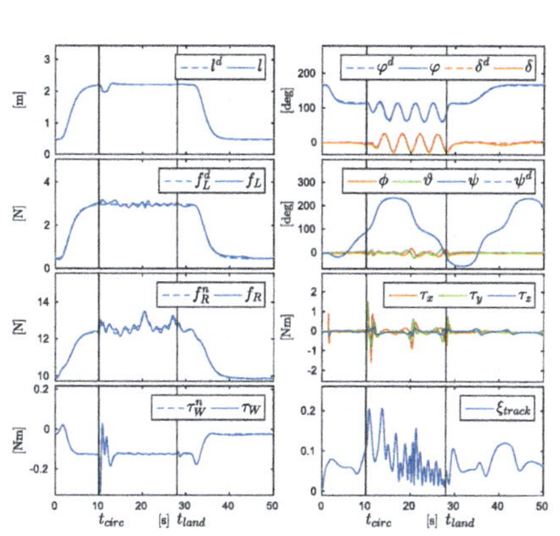

(a) Controller performances.

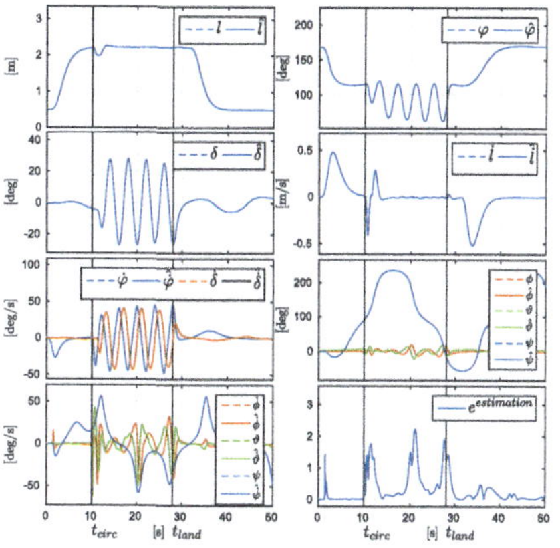

(b) Observer performances.

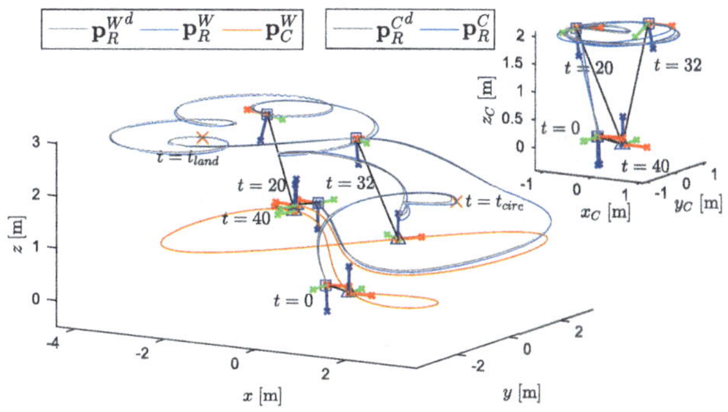

(c) Trajectories visualization.

Fig. 5.13 Simulation: non zero offset between tether attachment and center of gravity of the aerial vehicle

(not only vertical) noticing that, within some reasonable bounds, the system remains stable and with acceptable tracking performances.

5.3.6 Nondiagonal Inertia Matrix

In the derivation of the model and of the controller as well, we assumed a diagonal inertia matrix. In this section we check the robustness of the method if the aerial vehicle has a non diagonal inertia matrix. In particular, in Fig. 5.14, we show the results for a test in which the real inertia matrix is

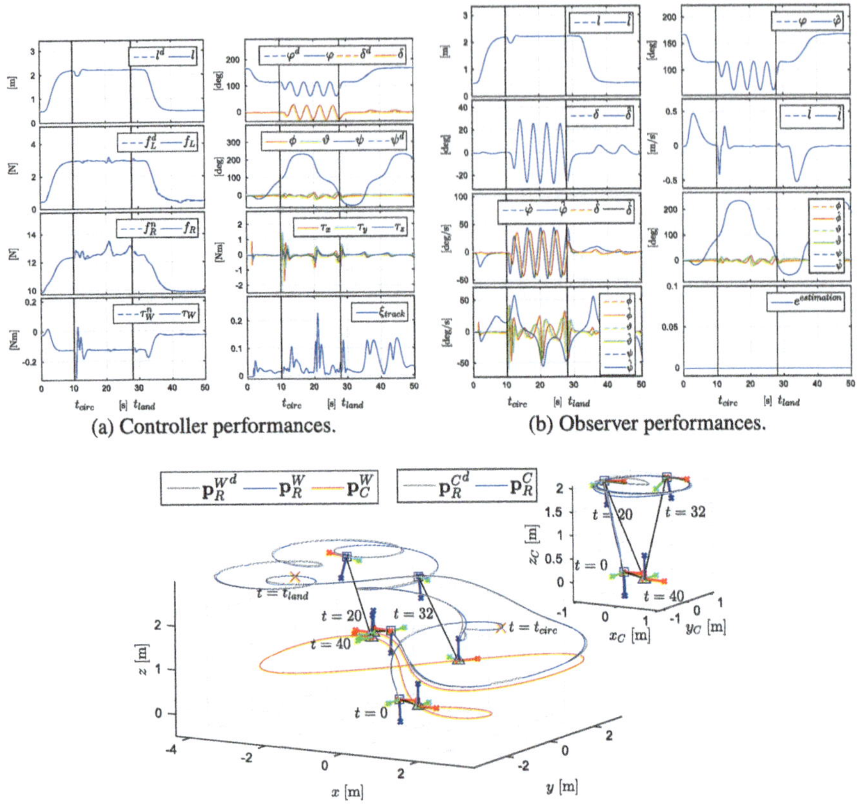

(a) Controller performances. (b) Observer performances.

(c) Trajectories visualization.

Fig. 5.14 Simulation: non-diagonal inertia matrix $\mathbf{J}_R$

$$\mathbf{J}_R = \begin{bmatrix} 0.25 & 0.05 & 0.05 \\ 0.05 & 0.25 & 0.05 \\ 0.05 & 0.05 & 0.25 \end{bmatrix},$$

while the controller still assumes a diagonal inertial matrix.

One can observe that the tracking error is not exactly zero but is kept limited within a small bound. For the observer this does not constitute a non ideality, in fact the estimation error is constantly zero.

With further simulations we observed that the system remains stable up to a value of 0.15 in the off diagonal terms (60% of the values on the main diagonal). With larger values the system becomes unstable.

5.3.7 Input Saturation

For how we planned the desired trajectory, the nominal input needed to track the desired outputs is always within the limits of the considered system. Indeed, exploiting the flatness, we are able to a priori check if the inputs exceed the minimum and maximum values. Nevertheless, in this section we want to show that the system is still stable if the inputs are hardly saturated for some instants. Thus we set some very restrictive limits on the input, i.e., $f_R \leq \bar{f}_R$ and $\underline{\tau} \leq \tau_i \leq \bar{\tau}$, where $i = x, y, z$, $\bar{f}_R = 13[\mathrm{N}]$, $\underline{\tau} = -1[\mathrm{Nm}]$ and $\bar{\tau} = 1[\mathrm{Nm}]$. In order to let the saturation show up during execution we did not re-plan the desired trajectory.

In Fig. 5.15a it can be seen that the inputs are saturated for some time instants during the execution of the task. When the inputs are saturated the tracking error increases, but, as soon as the inputs come back within the limits, the error exponentially decreases to zero.

We stress again the fact that the saturation of the inputs can be avoided exploiting the flatness. Using the flatness one can check if the desired trajectory requires inputs that are too large. In the worst case one can re-plan the trajectory such that the input limits are respected.

5.3.8 Motor Time Constant

With this simulation we want to further enlarge the set of non ideal models considered for the testing of the proposed control method. Considering an aerial vehicle actuated by rotating propellers, in this simulation we add the dynamical model of the motors described with a first order system characterized by a time constant of $\tau_M = 0.1[\mathrm{s}]$. In practice the propeller dynamics inserts a frequency dependent phase shift between the commanded control input and the actuated one, whose amplitude depends on the time constant. In other words, the models acts as a low pass filter on the commanded input, cutting its high frequency components. Those effects could dramatically decrease the performances or even make the system unstable. However, from Fig. 5.16, one can notice that our method is robust to the unmodeled effects of the propellers dynamics. Indeed, in some instant, where the trajectory is more dynamical and requires fast varying inputs, the tracking error increases but it is always bounded and at steady state converges to zero.

We remark that, if needed, one can increase the smoothness of the control inputs considering an higher order in the dynamic feedback control. Indeed adding more integrators on the control channels one can increase the degree of smoothness of the control input thus guarantying that it is always below the cutting frequency proper of the system, and in particular of the propellers. Another possible strategy is to exploit the flatness to plan a trajectory that fulfills the system limitations.

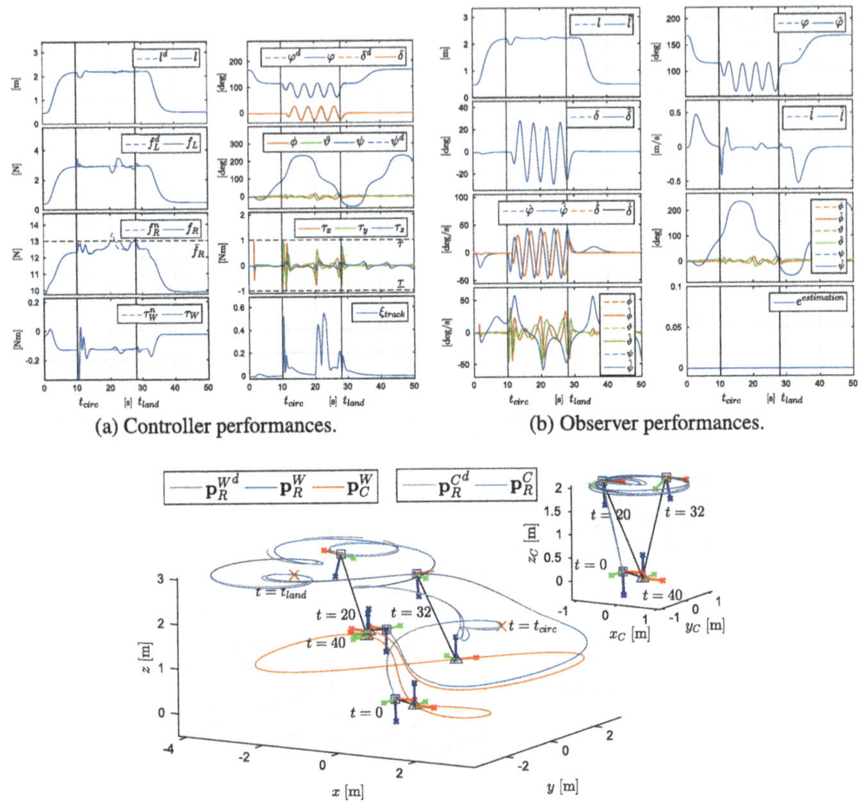

(c) Trajectories visualization.

Fig. 5.15 Simulation: saturation of the input

5.3.9 *Hierarchical Control Versus DFL Control for* $\mathbf{y}^a$

As we noted in Sect. 5.2, the hierarchical controller guarantees sufficiently good tracking performance in quasi static conditions. However the performance gets worse when the desired velocities and accelerations increase. In this section we shall compare the hierarchical controller tracking performance with respect to the one provided by the dynamic feedback linearizing controller. In particular, we simulate the system with Γ^a_{HC} together with the observer Γ^a_{HC}, tracking the same desired trajectory $\mathbf{y}^{ad}(t)$ used in Sect. 5.3, with an initial tracking error and under noisy estimated state. We shall then compare the results with the ones obtained with Γ^a_{DFL} tracking the same trajectory, and in particular under the non ideal cases (a) and (c) of Sect. 5.3 (see Fig. 5.7).

In Fig. 5.17 the results of the hierarchical controller in *ideal conditions* are reported. The initial tracking and estimation errors are the ones of case (a) in Sect. 5.3. After a tuning phase we were able to get a good performance and a small bounded

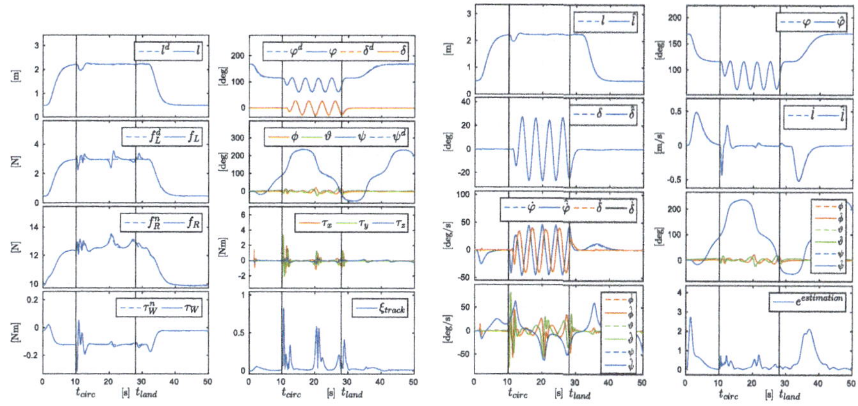

(a) Controller performances. (b) Observer performances.

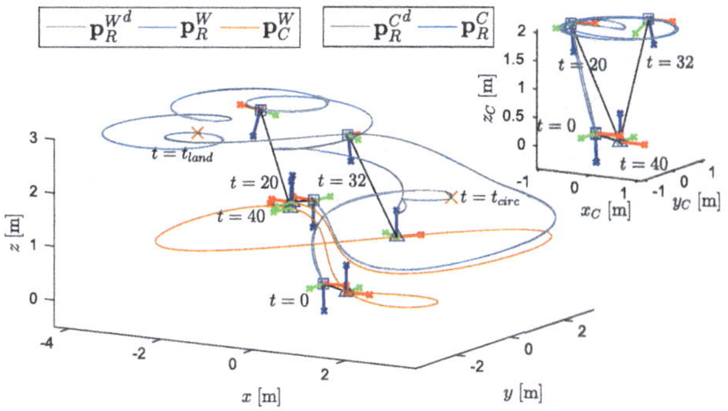

(c) Trajectories visualization.

Fig. 5.16 Simulation: system with motors dynamics

tracking error, even if the error does not converge exactly to zero. On the other hand, in the same conditions the controller based on dynamic feedback linearization is able to steer the output along the desired trajectory with zero error (see Fig. 5.7 or Sect. 5.3.1 for more details). However, to obtain good tracking performance with the hierarchical controller we had to set *very high* gains that make the system more reactive and thus able to follow the desired trajectory. Nevertheless this requirement has two main drawbacks.

The first drawback is that, due to the large control gains, the control effort increases thus possibly requiring an input that is out of the physical limits of the actuators. Indeed with this configuration we reach a maximum thrust and a maximum torque of about 15[N] and 2.5[Nm] respectively. This values are higher than the nominal inputs required to track the desired trajectory.

The second extremely serious issue arises in the presence of noise in the measurements and so in the estimated state. Indeed, the higher the gains, the larger the noise in the commands and the closer the controlled system is to instability. In fact,

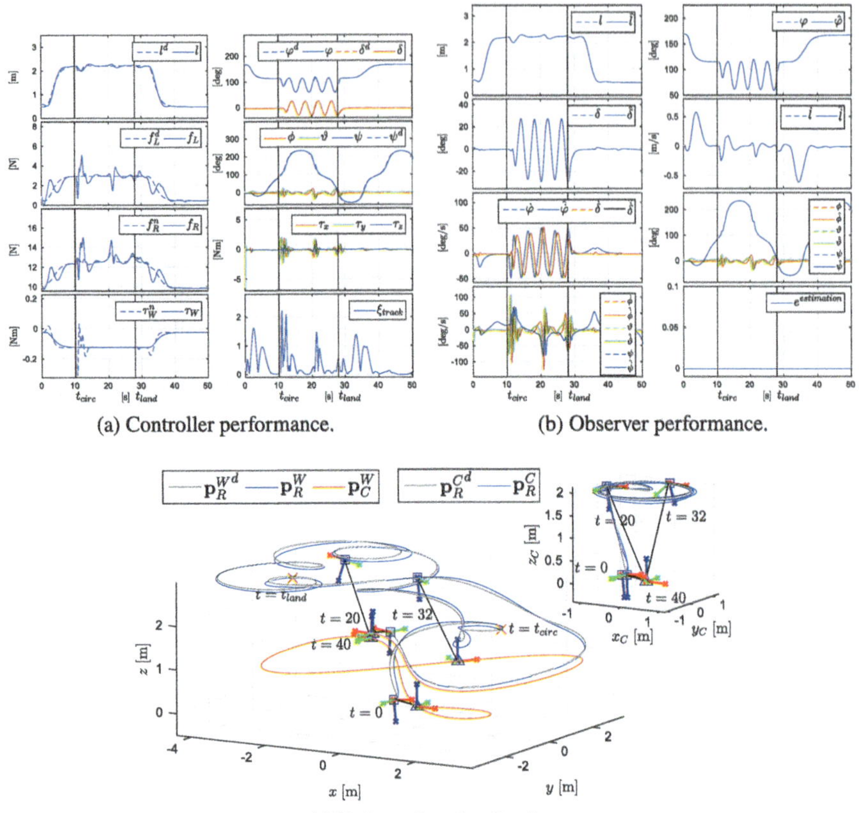

(a) Controller performance.

(b) Observer performance.

(c) Trajectories visualization.

Fig. 5.17 Simulation results: hierarchical control in ideal conditions with initial tracking error

simulating the system with the same measurement noise described in Sect. 5.3 (see Sect. 5.3.4 for more details) the closed loop system becomes unstable. In order to get a stable behavior we had to significantly lower the gains, an action that, however, clearly degrades the tracking performance. As we can see in Fig. 5.18 the performance with noise is much worse than the one obtained using the dynamic feedback linearizing controller in the same noisy condition.

Therefore, the hierarchical approach presents a strictly penalizing trade-off between applicability with noise and tracking performance. One cannot obtain both. Attainment of both objectives is instead possible with the DFL controller Γ^a_{DFL}. Nevertheless, we experimentally proven that, in standard conditions, with not too dynamic desired trajectories, the hierarchical controller can still guarantees good tracking performance with a minimum implementation effort. This controller allowed us to perform the landing and takeoff maneuvers on surfaced inclined up to 60°, in a very robust and reliable way. On the other hand, although the dynamic feedback linearization control provide much better performance, even in non ideal conditions,

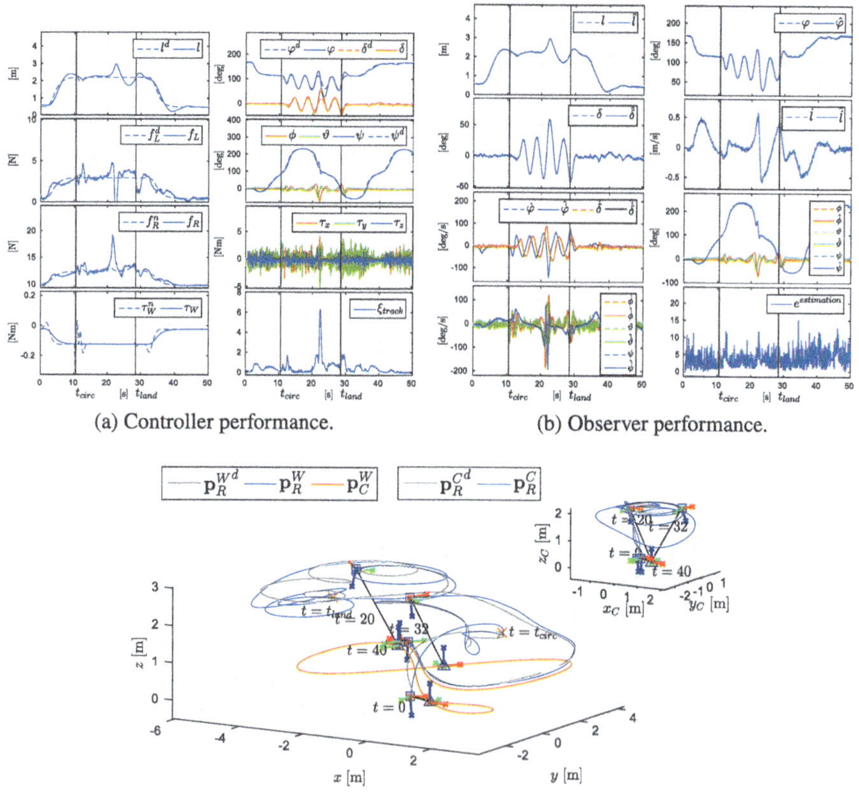

(a) Controller performance.

(b) Observer performance.

(c) Trajectories visualization.

Fig. 5.18 Simulation results: hierarchical control in the noisy case. To preserve stability lower gains have to be used with noise, therefore the performance is significantly degraded. The hierarchical controller presents a strictly penalizing trade off between tracking performance and robustness to noise

it comes with an higher computational cost, that makes it also more difficult to be implemented on a real robot.

5.3.10 DFL-Controller for $\mathbf{y}^c$ *in Case of Passive Link Actuator*

Recalling the discussion in Sect. 4.4.3.1, in order to obtain a steady state internal force $f_L{}^\star = 5\,[\mathrm{N}]$, we set the torque winch $\tau_W = -1\,[N]$. To obtain a sufficiently fast exponentially tracking, we set the controller gains $\mathbf{k}_i$ and $\mathbf{k}_4$ such that the error dynamics ξ_i and ξ_4 have poles in $(-0.5, -1, -1.5, -2)$ and $(-0.5, -1)$, respectively, for $i = 1, 2, 3$. Since the observer has been already tested in the previous section, here the control loop is closed with a direct measure of the state.

As before, we design the platform motion and the aerial vehicle desired trajectory in order to simulate a patrol-like task of a delimited area. The platform simply follows a certain trajectory shown in Fig. 5.19. The aerial vehicle, after the takeoff maneuver, at time t_{circ} has to loiter above the platform. Then, starting from time t_{land}, the aerial vehicle has to land on the platform.

To validate the control method and to test its robustness we performed several simulations in different non ideal conditions:

(a) We initialized the system with an *initial tracking error* of 10 [°] for the elevation, of 5 [°] for the azimuth and of 0.5 [m] for the link length. Looking at Fig. 5.19a we can notice that after a transient, the controller steers the output of interest along the desired trajectory. Notice that the internal force along the link remains always positive and close to the desired steady state value $f_L{}^\star$. Furthermore it is exactly equal to $f_L{}^\star$ whenever $\ddot{l}$ is zero.

(b) We tested the robustness of the control method with a *variation of the 5% on all the model's parameters* (see Fig. 5.19b). Due to the mismatch between real and nominal model, the feedback linearization is not exact and the error does not go to zero. However it remains always bounded showing nicely degrading and sufficiently good tracking performance. Moreover, in order to eliminate the constant error at steady state we have seen that a simple integral term in the linear control loop is sufficient. With further simulations we noticed that the system remains stable showing acceptable tracking errors up to a parametric variation of 50%, proving the robustness of the proposed method. Above the system becomes unstable.

(c) Although the control loop is not closed with the observer, we tested the robustness of the proposed method injecting *Gaussian noise on the measured state* used to close the control loop. The power of the noise has the same value of the one noticed in Sect. 5.3 out of the observer based on noisy sensors. From Fig. 5.19c one can see that the error does not converge to zero but remains always bounded showing good and practically viable tracking performance.

(d) In this simulation we considered the thrust and the torque of the aerial vehicle generated by *non ideal motors* modeled as a first order system characterized by a time constant of 0.2 [s]. The results displayed in Fig. 5.19d show a very small tracking error, validating the robustness of the control method to this additional non ideality.

5.4 Observer Based DFL-Controllers for Reduced Model

In this section we validate the observer based on IMU only, together with the 2D version of controllers Γ^a_{DFL} and Γ^b_{DFL} (see [8] for the corresponding details). In the first subsection we show the capability of Γ^a_{DFL} of independently controlling φ and

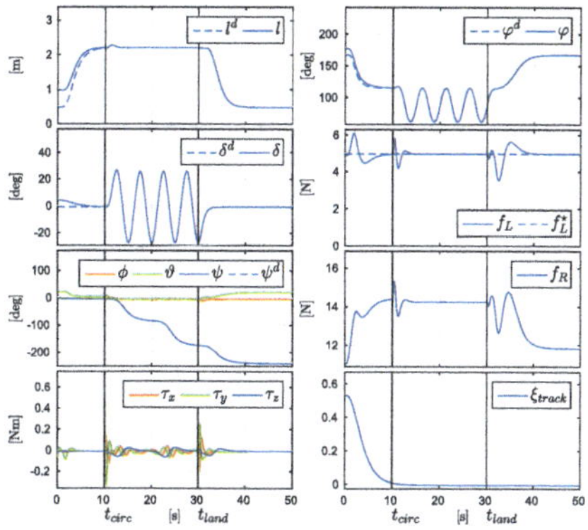

(a) Initial tracking error.

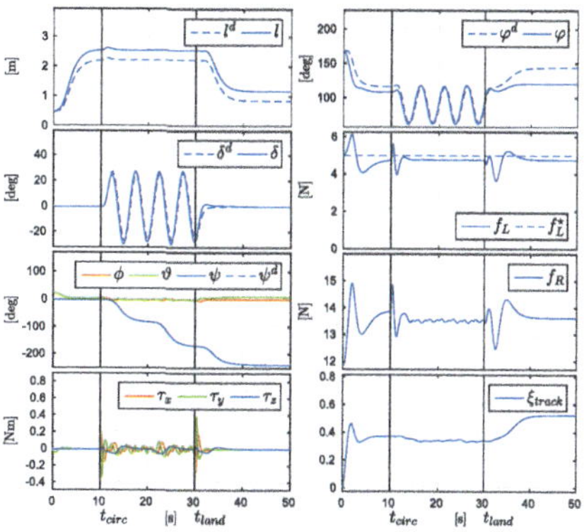

(b) Parameters uncertainties.

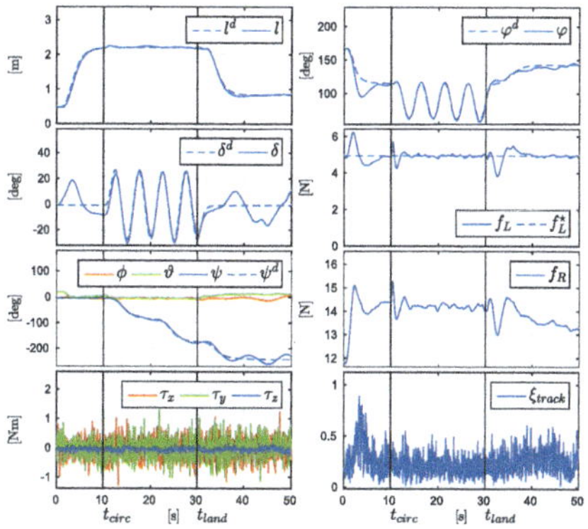

(c) Additional noise on the state.

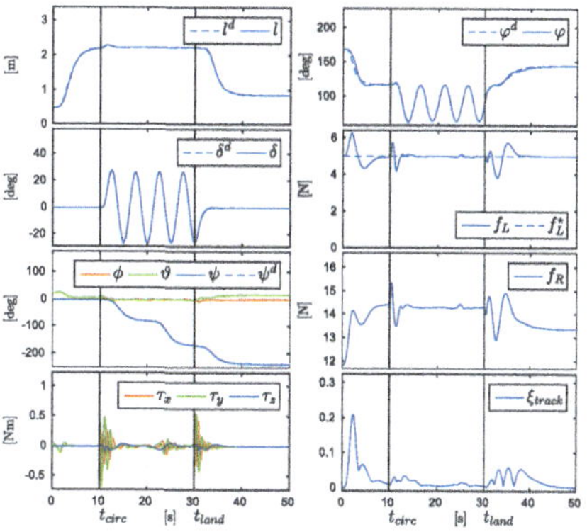

(d) Non ideal motors.

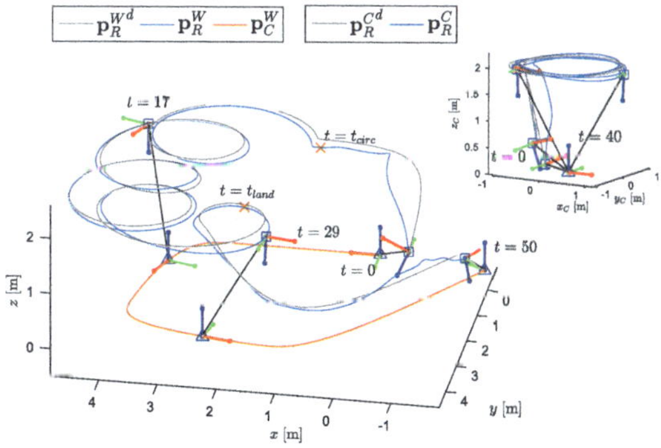

(e) Trajectories visualization for the case c).

Fig. 5.19 Simulation results: plausible task trajectory for the case of a passive link actuator. The performance for each non-ideal case are shown

f_L, even when the desired internal force trajectory goes from tension to compression, and vice versa. In the second subsection we instead provide a thorough analysis of the robustness of the 2D version of Γ^a_{DFL} and Γ^b_{DFL} together with the observe against non ideal conditions.

5.4.1 Controlling f_L for Both Tension and Compression

Figures 5.20a, b and 5.21a, b show the behavior of the system following smooth trajectories from an initial to a final output configuration. The plots of the tracking errors show that the proposed controller is able, after a short transient, to perfectly follow the time varying smooth trajectories of class C^3 and C^1 for the elevation and link internal force, respectively. An animation of the presented simulations is available at [9]. Notice that in Fig. 5.21b, the pick of torque at around time 2.6[s], arises due to the crossing of the control singularity, i.e., zero thrust. However, since that singularity is crossed only for one instant, the system remains stable.

We also tested controller Γ^a_{DFL} to track a desired trajectory of f_L that goes from an initial tension to a final compression, while following a desired elevation motion as well. In Fig. 5.21a it is interesting to notice that to pass from tension to compression the vehicle turns upside-down keeping the thrust always positive. On the other side, in the simulation of Fig. 5.21b the transition from tension to compression is obtained with the thrust that passes through zero and inverts its sign in order to obtain the same final compressing force of the simulation of Fig. 5.21a. This happens because in the second case the desired trajectory requires zero thrust at a certain moment. Then, since it is not possible to instantaneously turn the vehicle, the controller inverts the sign of the thrust in order to provide compression. In the case of vehicles able to provide also negative thrust this is not a problem. While, in the case of robots providing only positive thrust a planning phase is needed in order to generate feasible trajectories.

5.4.2 Robustness Investigation Against Non Ideal Conditions

In this section we present a comprehensive analysis of the robustness of the designed dynamic feedback linearizing controllers together with the observer based on IMU only, against non-ideal conditions. This shows both their strengths and possible limits when applied on a real system. In order to test the observer based only on the IMU measurements, the following analysis is carried out for the reduced model presented in Sect. 4.3, consisting of an aerial vehicle constrained on a 2D vertical plane tethered to a fixed point on the ground by a link with a constant length $l = 2$[m].

For this system we tested the reduced version of Γ^a_{DFL} and Γ^b_{DFL} together with the observer based only on the IMU measurement. We recall that the details of such reduced version of the presented DFL controllers can be found in [8].

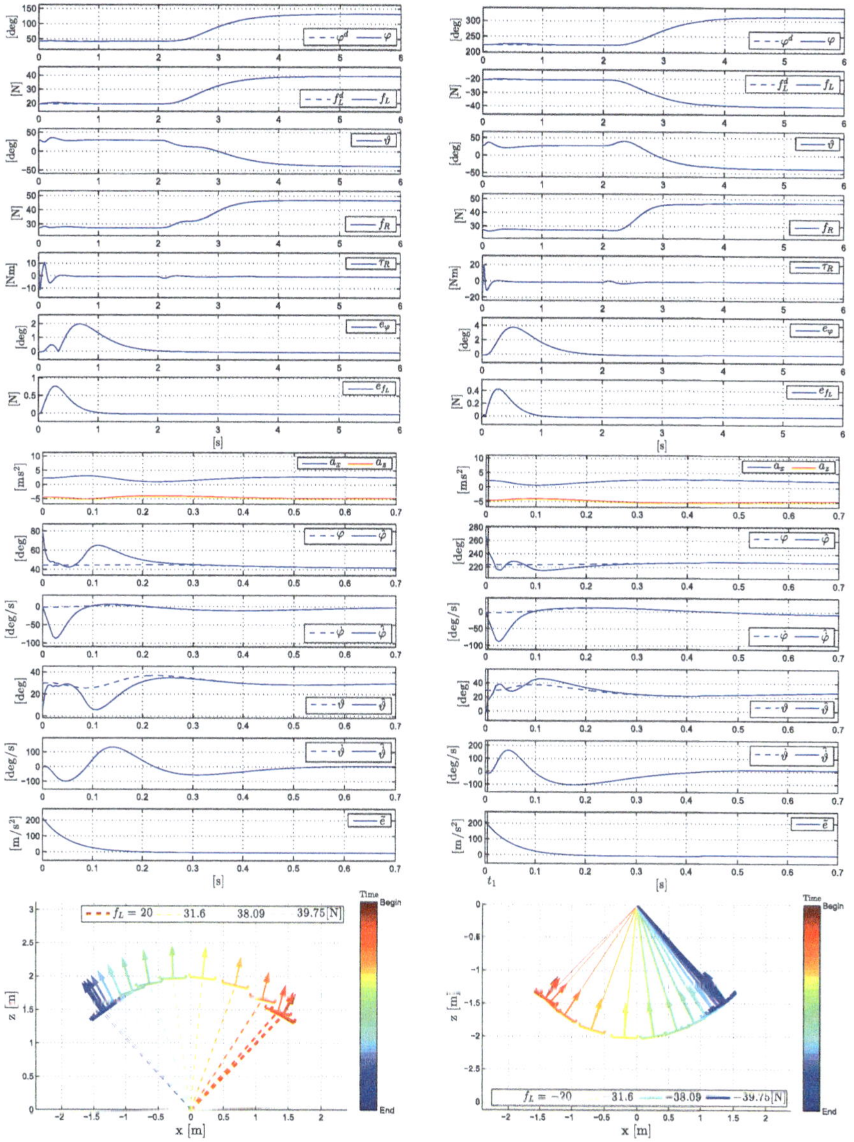

(a) The desired φ trajectories goes from the initial value $\varphi_0^d = \pi/4$ [rad] to the final $\varphi_f^d = 3\pi/4$ [rad]. The desired f_L trajectory goes from the initial tension of ${f_L}^d{}_0 = 20$ [N] to the final tension of ${f_L}^d{}_f = 40$ [N].

(b) The desired φ trajectories goes from the initial value $\varphi_0^d = 5\pi/4$ [rad] to the final $\varphi_f^d = 7\pi/4$ [rad]. The desired f_L trajectory goes from the initial compression of ${f_L}^d{}_0 = -20$ [N] to the final compression of ${f_L}^d{}_f = -40$ [N].

Fig. 5.20 Simulation results: controlling φ and the tension (on the left) or the compression (on the right)

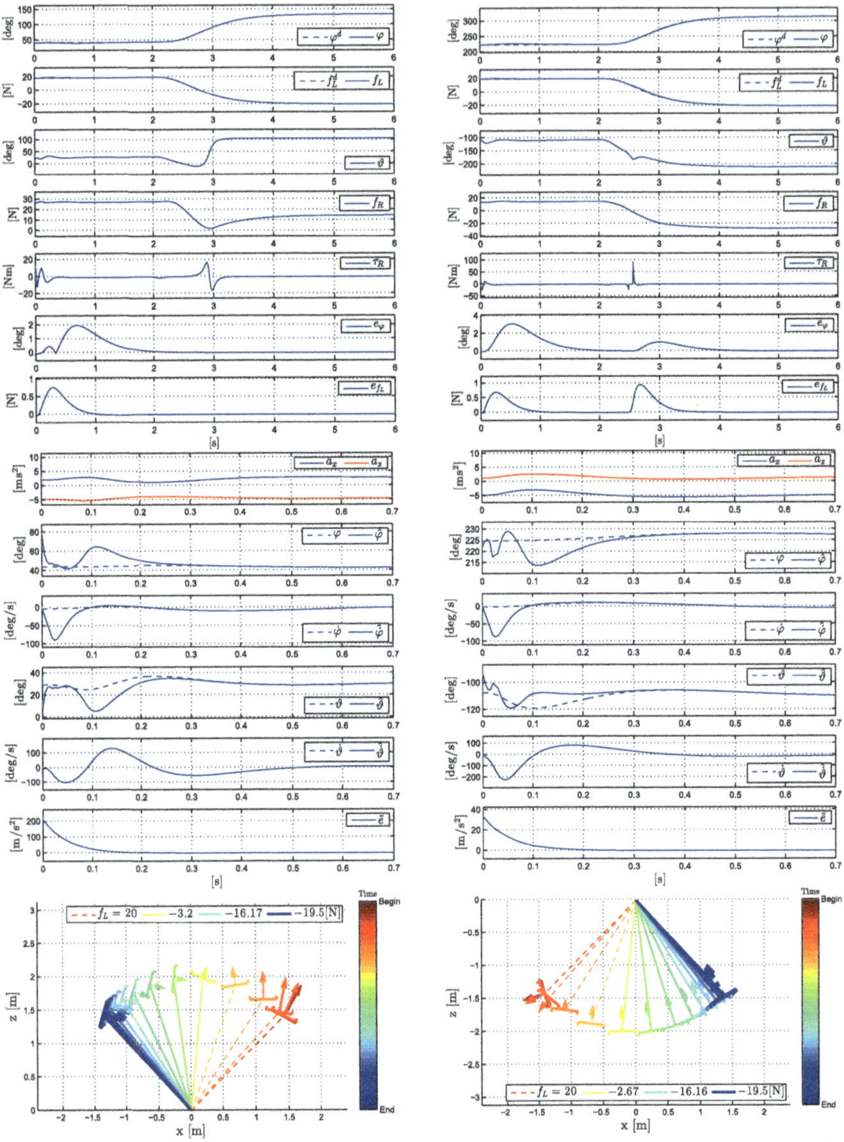

(a) The desired φ trajectories goes from the initial value $\varphi_0^d = \pi/4$ [rad] to the final $\varphi_f^d = 3\pi/4$ [rad]. The desired f_L trajectory goes from the initial compression of ${f_L}^d{}_0 = 20$ [N] to the final compression of ${f_L}^d{}_f = -20$ [N].

(b) The desired φ trajectories goes from the initial value $\varphi_0^d = 5\pi/4$ [rad] to the final $\varphi_f^d = 7\pi/4$ [rad]. The desired f_L trajectory goes from the initial tension of ${f_L}^d{}_0 = 20$ [N] to the final tension of ${f_L}^d{}_f = -20$ [N].

Fig. 5.21 Simulation results: controlling φ and f_L going from tension to compression. In figure **(b)**

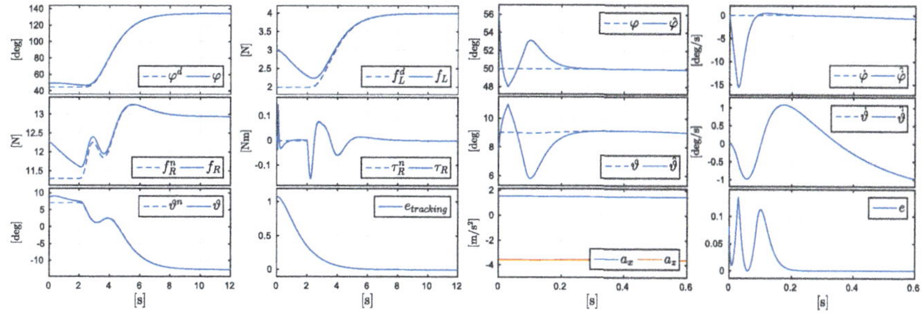

(a) Results of $\Gamma_{\mathrm{DFL}}^{a}$ and of the observer, on the left and on the right, respectively.

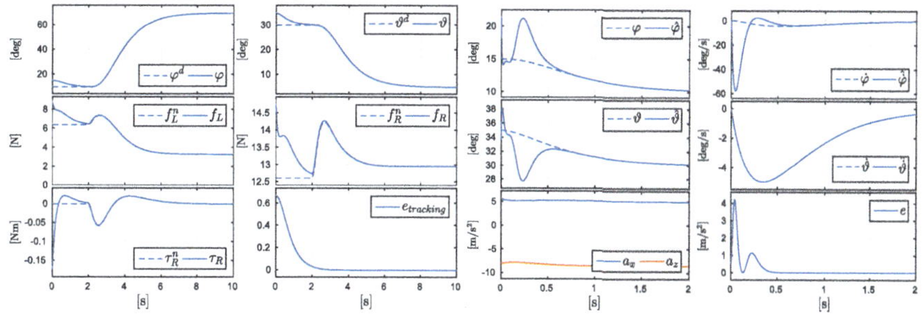

(b) Results of $\Gamma_{\mathrm{DFL}}^{b}$ and of the observer, on the left and on the right, respectively.

Fig. 5.22 Simulation results: nonzero initial tracking error

Concerning the controller $\Gamma_{\mathrm{DFL}}^{a}$ we set the gains of the linear outer control loop, $\mathbf{k}_1^a$ and $\mathbf{k}_2^a$, such that the error dynamics of φ and f_L has poles in $(-1, -1.5, -2, -2.5)$ and $(-1, -1.5)$ respectively. While for the controller $\Gamma_{\mathrm{DFL}}^{b}$ we set the gains $\mathbf{k}_1^b$ and $\mathbf{k}_2^b$, such that the error dynamics of φ and ϑ_A has poles in $(-0.5, -1, -1.5)$ and $(-0.5, -1)$ respectively. For the gains of the observer we set $\epsilon = 0.1$ and $(\alpha_1, \alpha_2, \alpha_3)$ such that the root of $s^3 + \alpha_1 s^2 + \alpha_2 s + \alpha_3$ are $(-6, -4.5, -3)$. Those values guarantee the stability of the closed loop system and a sufficiently rapid convergence of the observer and controller.

For the controller $\Gamma_{\mathrm{DFL}}^{a}$, the desired trajectory is a smooth step, continuous up to the fourth order for φ and up to the second order for f_L, from the initial values $\varphi_0^d = 45°$, ${f_L}_0^d = 3$[N], to the final values $\varphi_f^d = 135°$, ${f_L}_f^d = 5$[N], respectively. Smooth step-like trajectories (see Fig. 5.22), as it will be clear later, have the benefit of clearly showing the performance of the controllers under three important conditions: the initial transient, the tracking of a fast time-varying signal, and the steady state. For the controller $\Gamma_{\mathrm{DFL}}^{b}$, the desired trajectory is a smooth step, continuous up to the third order for φ and up to the second order for θ, from the initial values $\varphi_0^d = 10°$, ${\vartheta_A}_0^d = 30°$ to the final values ${\vartheta_A}_f^d = 50°$, ${\vartheta_A}_f^d = 5°$, respectively. Notice that, since the system is constrained to the 2D vertical plane with $\mathbf{y}_R = \mathbf{y}_W$, we have that $\vartheta_A = \vartheta$, where we recall that ϑ is the pitch of the vehicle.

To obtain a complete validation, in the following we show a concise summary of the results about the stability and robustness of the proposed method under different non-ideal conditions, such as: (a) nonzero initial tracking and estimation errors, (b) parametric variations, (c) generic CoM position and non-negligible link mass, (d) noisy sensor measurements, and (e) non-ideal motors.

5.4.2.1 Validation for Nonzero Initial Tracking/Estimation Errors

In order to show the asymptotic convergence performance of both the controller and the observer we initialize the control system with nonzero initial tracking and estimation errors. One can see in Fig. 5.22 that, after the convergence of the observer, which takes less than one second, the controller Γ^a_{DFL} is able to steer the outputs along the desired trajectories with zero error. A similar behavior is obtained for the controller Γ^b_{DFL}. We then performed many other similar simulations with different initial errors and we observed always the same asymptotically convergent behavior, as expected from the almost-global nature of the proposed observer and control laws.

5.4.2.2 Parametric Variations

We notice that in principle one could try to design an adaptive control law that is able to compensate for parametric uncertainties. However, this is clearly a tough objective, because the system is nonlinear and the available measurements are only the (nonlinear) accelerometer and the gyroscope readings. Therefore this goal is left as future work. Instead, we concentrate in this section on assessing the ranges of parameter variations that causes a degradation of the performance that remains within an acceptable bound. By doing so, we shall see in fact that the proposed control scheme possesses a remarkable robustness even without the presence of an adaptive design.

Considering l_0, m_{R0} and J_{R0} the real parameters value and l, m_R and J_R the nominal ones, we set $l = (1 + \Delta l)l_0$, $m_R = (1 + \Delta m_R)m_{R0}$ and $J_R = (1 + \Delta J_R)J_{R0}$, where Δm_R, Δl and ΔJ_R denote the corresponding parametric variations.

For obtaining a comprehensive analysis we tested the behavior for several different parametric variation combinations. The results are plotted in Figs. 5.23 and 5.24, where we show the mean tracking error, $\bar{e}_{track}$, and the corresponding standard deviation $\sigma_{\bar{e}_{track}}$, defined as

$$e_{track}(t) = \frac{\|y_1^d(t) - y_1(t)\|}{y_1^d(t)} + \frac{\|y_2^d(t) - y_2(t)\|}{y_2^d(t)}$$

$$\bar{e}_{track} = \frac{1}{t_f - t_0} \int_{t_0}^{t_f} e_{track}(t)dt$$

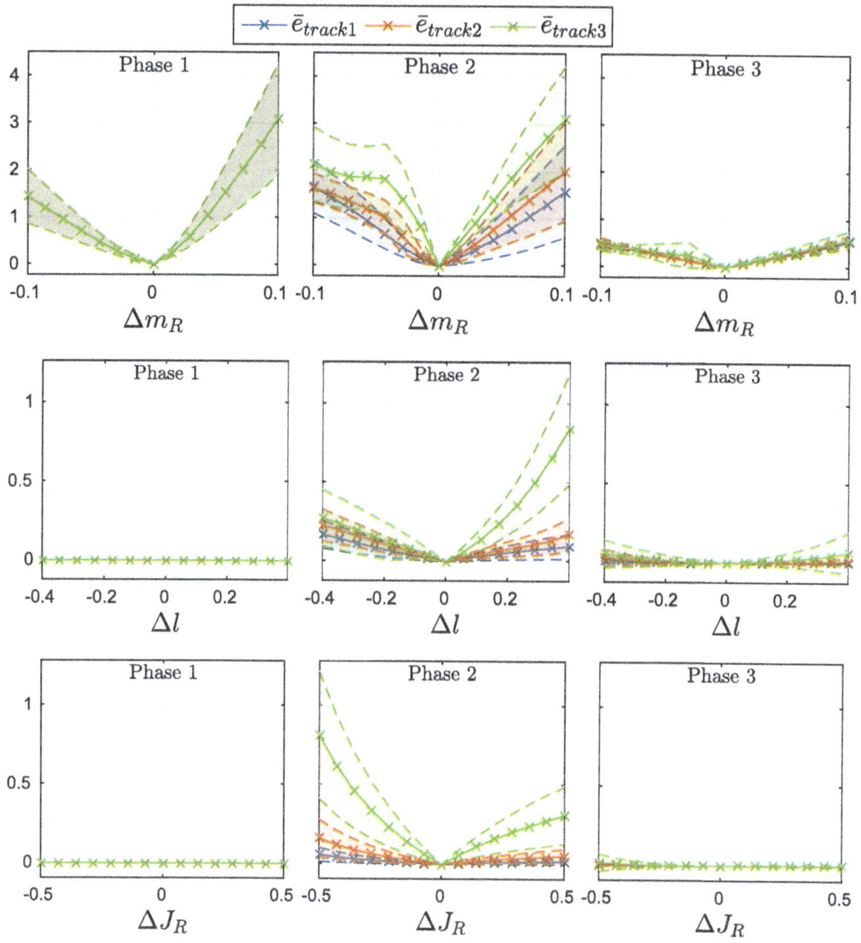

Fig. 5.23 Simulation results: parametric variation—Controller Γ^a_{DFL} (elevation and link force). The subscript 1, 2, and 3 correspond to the three different trajectory times. Outside of the displayed range of parametric variation the performance is unacceptable or the closed loop system results to be even unstable. © 2020 IEEE. Reprinted, with permission, from [8]

$$\sigma_{\bar{e}_{track}} = \sqrt{\frac{1}{t_f - t_0} \int_{t_0}^{t_f} (e_{track}(t) - \bar{e}_{track})^2 dt},$$

where t_0 and t_f are the initial and final time, respectively. Notice that for the reduced model $y_1 = \varphi$ and $y_2 = f_L$ or $y_2 = \vartheta_A$ for Γ^a_{DFL} and Γ^b_{DFL} respectively. In the plots the solid line corresponds at the mean tracking error while the dashed lines correspond at the mean tracking error plus and minus its standard deviation.

The effect of an unknown parameter could also change with respect to the trajectory and in particular with respect to the velocity and acceleration at which the

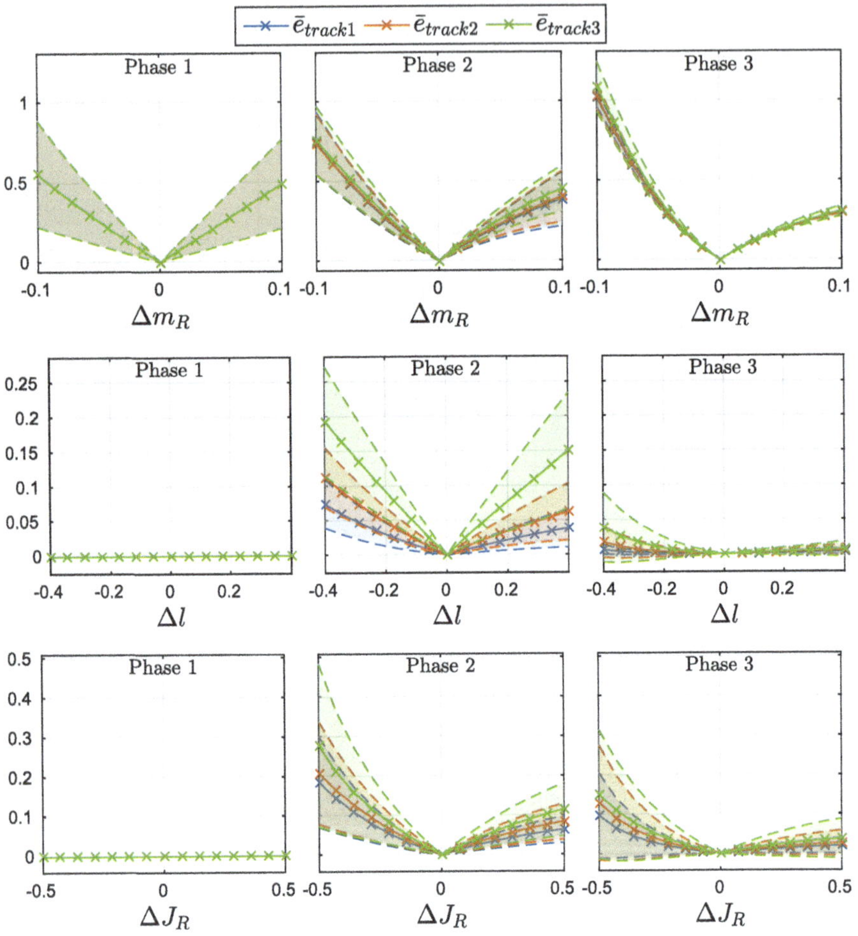

Fig. 5.24 Simulation results: parametric variation—Controller Γ^b_{DFL}(elevation and attitude). © 2020 IEEE. Reprinted, with permission, from [8]

path is followed. Consequently we plotted the mean tracking error, $\bar{e}_{track1}$, $\bar{e}_{track2}$ and $\bar{e}_{track3}$, for the same type of desired path (smooth step) but executed with three different durations (increasing velocity): (1) 7[s], (2) 5[s] and (3) 3[s] respectively. We also analyze the error behavior dividing the trajectory into three phases: in the *Phase 1* (transient) the desired trajectory is constant and the analysis is more focused on the convergence of the observer; *Phase 2* constitutes the dynamic part where the desired trajectory quickly goes from the initial value to the final one; the *Phase 3*, the last, corresponds to the steady state condition where the desired trajectory is again constant. We show the tracking error for each of the three phases to better understand if a parameter variation affects more the transient, the dynamic phase, or the static one.

From Figs. 5.23 and 5.24 one can notice that, as expected, the performance gets worse increasing the parametric variation. Furthermore, the same variation has more effect if the trajectory is more "aggressive" and it is followed with higher speed ($\bar{e}_{track3}$). This is due to the fact that with higher speed and acceleration the inertial and Coriolis/centripetal terms become larger, and thus also the error in the feedback linearization increases, which in turn implies a worst tracking.

Comparing the performance between the two controllers, we notice that controller Γ^b_{DFL} results to be more robust than controller Γ^a_{DFL} in term of mean tracking error. This is due to the fact that for the controller Γ^b_{DFL}, the dynamics of one of the controlled outputs, namely ϑ_A, is not influenced by the parameters such as mass and length of the link. This means that any variation on these parameters does not generates a worse tracking of ϑ^d_A, which results in a lower tracking error.

One can also notice that the mean tracking error is not in general symmetric with respect to the sign of the corresponding parametric variation. For example for the controller Γ^b_{DFL} it is better to overestimate the mass, and the length of the link rather than underestimating them, while for controller Γ^a_{DFL} it results to be the opposite, even if these consideration are more relevant for the dynamic phase. Indeed, during the steady state phase the behavior is almost symmetrical.

Another fact that appears clear from the plots is that the variation that most influences the performance is the one on the mass of the aerial vehicle. Fortunately, in practice this parameter can be easily measured with high precision.

5.4.2.3 Generic CoM Position and Non-negligible Link Mass

The controllers developed in this paper assume that the system can be represented with the model given in Sect. 4.3, where the CoM of the aerial vehicle coincides with the attachment point of the link to the vehicle and the link has a negligible mass. Figure 5.25 represents instead a more general model, for which the assumptions done in Sect. 4.3 are not fulfilled. Taking into account the definitions made in Sect. 4.3 we then define a *body frame*, $\mathcal{F}_l$, attached to the link, with axes $\{\mathbf{x}_l, \mathbf{y}_l, \mathbf{z}_l\}$ and origin O_l coinciding with the center of mass (CoM) of the link. The position of O_l, defined in $\mathcal{F}_W$, is denoted with $\mathbf{p}_l = [x_l \;\; y_l \;\; z_l]^\top$. As for $\mathcal{F}_R$ we have that $\mathbf{y}_l \equiv \mathbf{y}_B \equiv \mathbf{y}_W$ and $y_l \equiv$

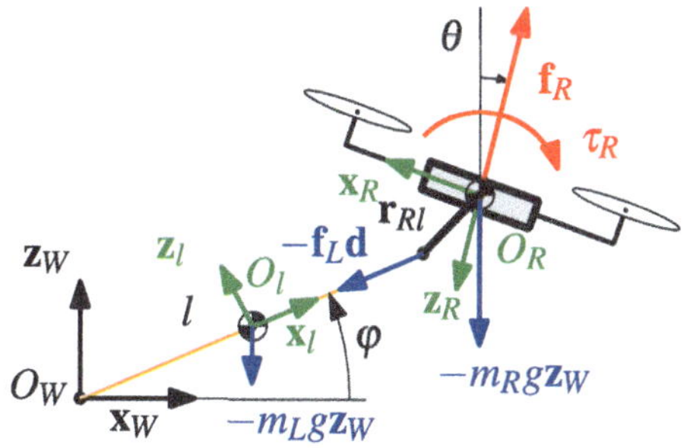

Fig. 5.25 Representation of the more general system and its variables, still constrained in the 2D vertical plane. © 2020 IEEE. Reprinted, with permission, from [8]

0. For the validation we model the link as a rigid body of mass $m_L \in \mathbb{R}_{>0}$ and inertia $J_L \in \mathbb{R}_{>0}$. Considering the inertia of the link as the inertia of an infinitesimally thin rigid tie with uniform distributed mass, we have also that $J_L = m_L l^2/12$. Assuming links with high stiffness, the deformations and the elongations results negligible with respect to the length of the cable itself, in the range of forces of our concern. Therefore the link length is fixed. The link is connected at one end to a fixed point coinciding with O_W and at the other end to a point rigidly attached to the aerial vehicle whose constant position in $\mathcal{F}_R$ is denoted with $\mathbf{r}_{Rl} = [r_x \;\; 0 \;\; r_z]^\top$. If $\|\mathbf{r}_{Rl}\| = 0$ then the link is directly attached to the CoM of the aerial vehicle.

The mechanical model of the more general robotic system can be then derived writing the dynamics as the one in (4.18) plus a disturbance due to the non idealities:

$$\mathbf{M}'(\mathbf{q}')\ddot{\mathbf{q}}' + \mathbf{g}'(\mathbf{q}') + \delta(\mathbf{q}', \dot{\mathbf{q}}', \ddot{\mathbf{q}}', \mathbf{u}') = \mathbf{Q}'(\mathbf{q}')\mathbf{u}',$$

where

$$\delta(\mathbf{q}', \dot{\mathbf{q}}', \ddot{\mathbf{q}}', \mathbf{u}') = \bar{\mathbf{M}}(\mathbf{q}')\ddot{\mathbf{q}}' + \bar{\mathbf{c}}(\mathbf{q}', \dot{\mathbf{q}}') + \bar{\mathbf{g}}(\mathbf{q}') - \bar{\mathbf{Q}}(\mathbf{q}')\mathbf{u}',$$

$$\bar{\mathbf{M}} = \begin{bmatrix} \bar{J}_\varphi & J_{\varphi\theta} \\ J_{\theta\varphi} & \bar{J}_\theta \end{bmatrix}, \quad \bar{\mathbf{c}} = \begin{bmatrix} \bar{c}\dot{\theta}^2 \\ \bar{c}\dot{\varphi}^2 \end{bmatrix},$$

$$\bar{\mathbf{g}} = \begin{bmatrix} \frac{m_L}{2} l g \mathbf{d}^\perp \cdot \mathbf{e}_3 \\ -m_R l g \bar{\mathbf{R}}_B^W \mathbf{r}_{Rl} \cdot \mathbf{e}_3 \end{bmatrix}, \quad \bar{\mathbf{Q}} = \begin{bmatrix} 0 & 0 \\ -\mathbf{r}_{Rl} \cdot \mathbf{e}_3 & 0 \end{bmatrix},$$

where $\bar{\mathbf{R}}_B^W = \partial \mathbf{R}_B^W/\partial\theta$, $\bar{J}_\varphi = m_L l^2/3$, $\bar{J}_\theta = m_R \|\mathbf{r}_{Rl}\|^2$, $J_{\varphi\theta} = J_{\theta\varphi} = -m_R l \bar{\mathbf{R}}_B^W \mathbf{r}_{Rl} \cdot \mathbf{d}^\perp$, $\bar{c} = m_R l \bar{\mathbf{R}}_B^W \mathbf{r}_{Rl} \cdot \mathbf{d}$.

For a plausible case in which the link consists of a cable of mass $m_L = 0.01 m_R$ and inertia (during taut condition) $J_L = m_L l^2/12$, and it is attached to the robot in the position $r_{BL} = [0.03 \;\; 0 \;\; 0.03]^\top$[m] with respect to $\mathcal{F}_R$, we noticed that the controlled system is stable but the error does not converge exactly to zero. Indeed, due to the nonzero $\|\mathbf{r}_{Rl}\|$, the force along the link generates an extra torque on the aerial vehicle that is not compensated and so a constant steady state error appears.

In order to understand how each parameter of the more general model affects the tracking performance, as before, we show in Fig. 5.26 the mean tracking error and its standard deviation for different parameter values and in the three phases described before. In particular the mass of the link is taken as $m_L = \Delta m_L m_R$.

We noticed that the negative effects due to a nonzero offset $\mathbf{r}_{BL}$ reduce or increase if the rotational inertia is increased or reduced, respectively. Indeed, looking at the rotational dynamics in the case of non zero offset:

$$\ddot{\vartheta} = \tau_R/J_R - f_L \mathbf{d} \cdot \mathbf{r}_{BL}/J_R, \tag{5.1}$$

one can notice that the effect of the link force on the angular acceleration decreases if the inertia increases. Intuitively, a bigger inertia would mean a bigger mass or a bigger dimension of the vehicle. In the second case, the bigger the vehicle the more

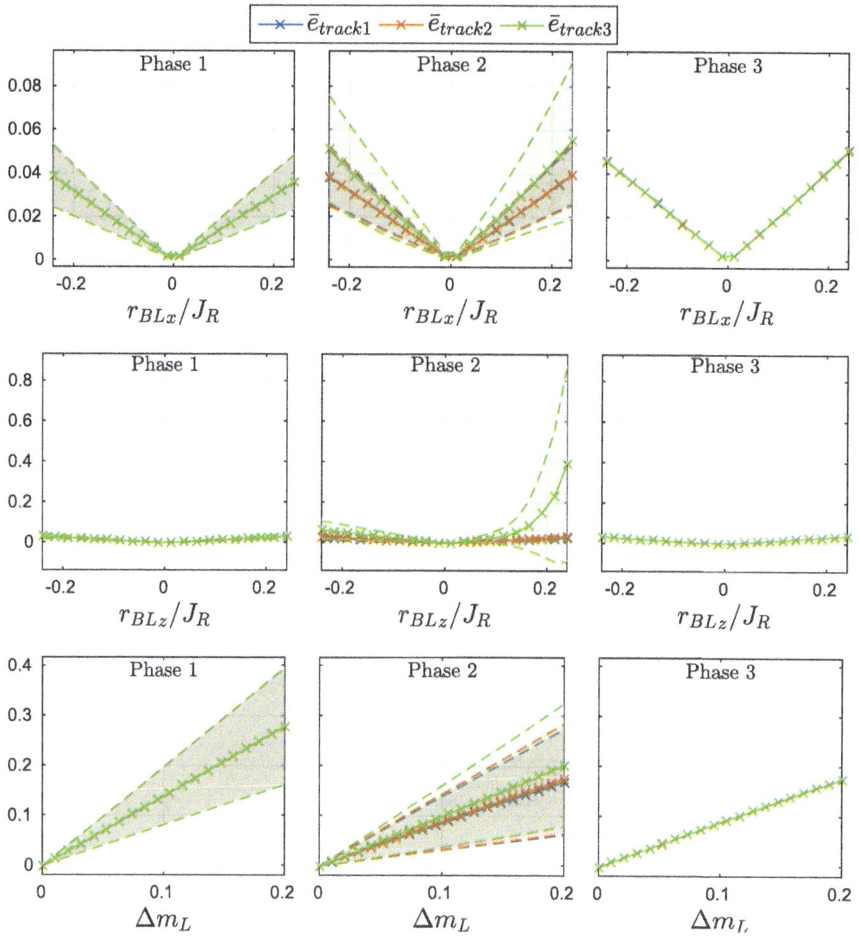

Fig. 5.26 Simulation results: controller Γ^a_{DFL}. Mean tracking error when changing the parameters of the general model.

the effect of the offset becomes negligible. For this reason in Fig. 5.26 we plot the mean tracking error with respect to $r_{BL_\star}/J_R$, thus normalizing this effect.

We did the same test for the controller Γ^b_{DFL}, which resulted to be much more sensitive to link mass and to the offset than the controller Γ^a_{DFL}. This is due to the fact that one of the output, namely the attitude of the aerial vehicle ϑ_A, is directly influenced by the offset as it is shown in (5.1) (we recall the in the 2D case $\vartheta_A = \vartheta$). Furthermore, even with a small offset the tracking error is such that the actual trajectory passes through the singularity of the controller Γ^b_{DFL} (see Sect. 4.6.2) causing an unstable behavior.

On the other hand, the controller Γ^a_{DFL} turned to be much more robust to these sort of structural model variations. From Fig. 5.26 we can see that the parameters that mostly affect an increase of the error are the entries of $\mathbf{r}_{BL}$, i.e., r_{BLx} and r_{BLz}. One can notice that it is more advisable to attach the link such that one is sure that $r_{BLz} \leq 0$, especially if agile motions are required. The effect of the displacement along $\boldsymbol{x}_R$ is instead almost symmetric. The small asymmetry is due to the particular trajectory passing from the first to the second quadrant. The mean tracking error increases instead almost linearly with respect to the mass of the link. Nevertheless, even with m_L equal to the 20% of m_R the closed loop system remains still perfectly stable.

5.4.2.4 Noisy Measurements

In this section we investigate the robustness of the proposed method in presence of noisy measurements, which always exist in reality. We consider both the accelerometer and the gyroscope measures being affected by a white Gaussian noise of variance $0.1[\text{m/s}^2]$ and $0.01[\text{rad/s}]$ respectively.

From Fig. 5.27 we can observe that the estimated state shows some noise but the corresponding error remains always bounded. Due to the noisy component on the estimated state the control action presents some oscillations that imply a non exact

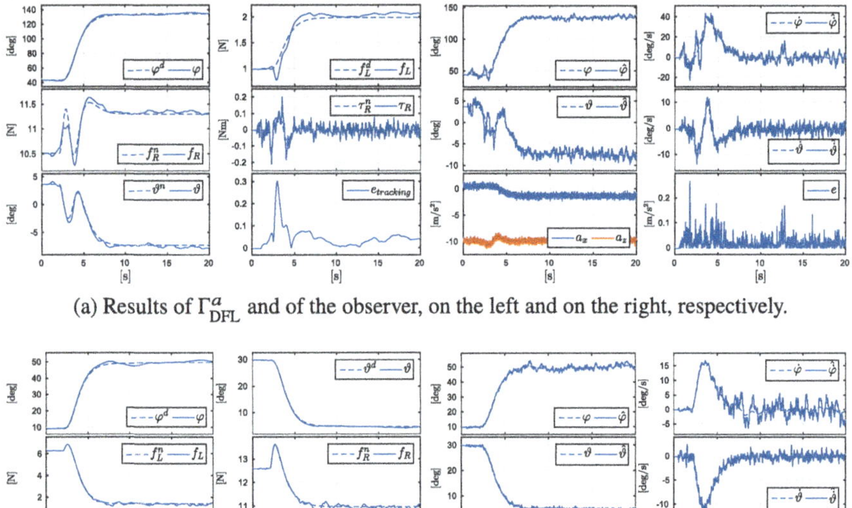

(a) Results of Γ^a_{DFL} and of the observer, on the left and on the right, respectively.

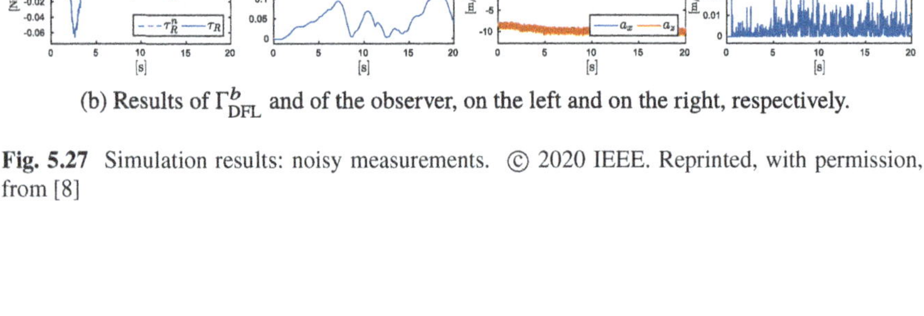

(b) Results of Γ^b_{DFL} and of the observer, on the left and on the right, respectively.

Fig. 5.27 Simulation results: noisy measurements. © 2020 IEEE. Reprinted, with permission, from [8]

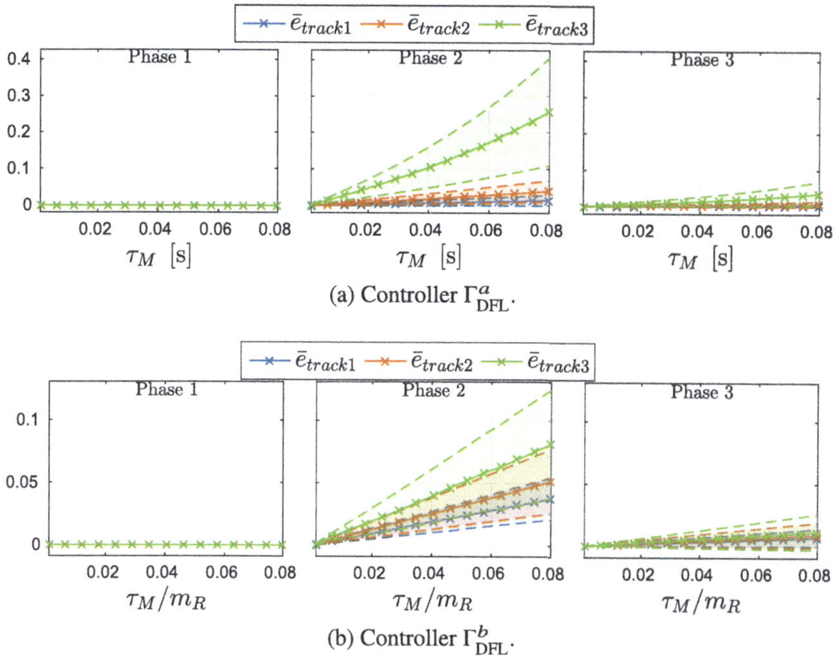

Fig. 5.28 Simulation results: non ideal motors. Mean tracking error when changing the motor time constant

tracking of the desired trajectory. Nevertheless the tracking error remains small and always bounded. Notice that to achieve these results we had to reduce the gains of both the controller and observer. Indeed, high gain values increase the convergence speed but also amplify the sensitivity to noisy measurements. In general the two controllers does not show particularly different behaviors in face of the presence of noise.

5.4.2.5 Non-ideal Motors

In a real scenario, one motor cannot instantaneously change the spinning velocity of the propeller, and in turn the thrust produced. Indeed, this discontinuous variation of the speed would require the application of an infinite torque by the motor, that is clearly not possible. Instead the dynamics of the motor is characterized by a certain time constant, $\tau_M \in \mathbb{R}$, that quantifies the time needed to change the motor speed. In this section we analyze this additional non ideality testing the proposed method with different non ideal motors characterized by an increasing time constant. In Fig. 5.28 we show the relative mean and variance of the tracking error for the different time constant values τ_M. The plots clearly shows that increasing the time constant the tracking error increases as well, especially during the dynamic part of the desired trajectory (Phase 2). Indeed, for motors with higher time constant, the error between commanded and actuated thrust on each propeller increases causing a bigger tracking error. However, the system remains stable up to a time constant of 0.08[s], which is completely acceptable in real systems.

This analysis is important for the scalability of the system. Indeed, bigger vehicles with higher mass imply the need of an higher lift that can be in general generated by bigger propellers. This in turn requires the use of bigger motors that are characterized by a larger time constant. Finally, as shown in Fig 5.28, the larger mass of the system, and so the larger time constant of the motors, reduces the tracking performance of the system for dynamic trajectories. Therefore, when we increase the dimension and the mass of the vehicle, in order to still obtain good tracking performance, it is necessary to reduce the agility of the desired maneuver reducing the demanded accelerations.

References

1. Franchi, A., Mallet, A.: Adaptive closed-loop speed control of BLDC motors with applications to multi-rotor aerial vehicles. In: 2017 IEEE International Conference on Robotics and Automation, pp. 5203–5208. Singapore (2017)
2. Kolmanovsky, I., Garone, E., Cairano, S.D.: Reference and command governors: a tutorial on their theory and automotive applications. In: 2014 American Control Conference, pp. 226–241. IEEE (2014)
3. Video 1. Takeoff and landing on slopes via inclined hovering with a tethered aerial robot (2016). https://www.youtube.com/watch?v=01UYN289YXk
4. Tognon, M., Dash, S.S., Franchi, A.: Observer-based control of position and tension for an aerial robot tethered to a moving platform. IEEE Robot. Autom. Lett. **1**(2), 732–737 (2016)
5. Video 2. Observer-based control of position and tension for an aerial robot tethered to a moving platform (2016). https://www.youtube.com/watch?v=VGnTBFCPbWg
6. Sandino, L., Santamaria, D., Bejar, M., Viguria, A., Kondak, K., Ollero, A.: Tether-guided landing of unmanned helicopters without GPS sensors. In: 2014 IEEE International Conference on Robotics and Automation, pp. 3096–3101. Hong Kong, China (2014)
7. Mahony, R., Hamel, T., Pflimlin, J.-M.: Nonlinear complementary filters on the special orthogonal group. IEEE Trans. Autom. Control **53**(5), 1203–1218 (2008)
8. Tognon, M., Franchi, A.: Dynamics, control, and estimation for aerial robots tethered by cables or bars. IEEE Trans. Robot. **33**(4), 834–845 (2017)
9. Video 3. Nonlinear observer-based tracking control of link stress and elevation for a tethered aerial robot using inertial-only measurements (2015). https://www.youtube.com/watch?v=6Y3NDnhLWr0

Chapter 6
Theory and Experiments for a Practical Usecase

6.1 The Problem of Landing and Takeoff On/from a Sloped Surface

In many aerial robot applications such as search and rescue, the task consists on providing assistance in hostile environments such as mountains or civil areas after natural catastrophes. In this scenarios it is very likely that the terrain is not flat, making the landing and takeoff maneuvers of the aerial robot very complicate and unsafe.

The problem of landing on a sloped (not flat) surface is a very challenging problem for an unidirectional-thrust aerial vehicle due to its underactuation. In fact, the task requires to control both position and attitude since the vehicle has to be oriented as the surface on which we want to land, but this is not possible. It is well known that one can control the position of an unidirectional-thrust aerial vehicle, but not its attitude. The latter is indeed a byproduct of the particular position trajectory that we want to follow (given by the differential flatness). The classical approach for free-flying vehicles is based on motion planning [1–3] (sometimes called *perching* maneuver). It consists on exploiting the flatness of the system with respect to the position [4] to plan a desired trajectory such that the vehicle ends the maneuver with the proper position and orientation. Different controllers can be then applied to track this trajectory. However, the success of the maneuver requires an almost perfect tracking that implies an almost perfect state estimation and knowledge of the model. Otherwise, small deviations from the nominal trajectory would lead to miss the target or to crash on it.

On the other hand, we shall show that the use of a tether is very useful to solve the faced problem of landing and takeoff on a sloped surface. As we saw in Chap. 4, for a tethered aerial vehicle we have the great advantage to partially control the attitude of the vehicle. Under certain conditions better stated in the following, this property

M. Tognon and A. Franchi, *Theory and Applications for Control of Aerial Robots in Physical Interaction Through Tethers*, Springer Tracts in Advanced Robotics 140,
https://doi.org/10.1007/978-3-030-48659-4_6

allows to perform the landing and takeoff maneuvers in a very reliable way, even in the presence of model errors, and for almost any sloped surface.

In Chaps. 4 and 5, we already showed the case of a tether aerial vehicle, together with an actuated link, landing and taking off on/from a flat moving surfaces. Nevertheless, this configuration requires to add an actuator that increases the complexity of the system and reduces its already limited payload if placed on-board. For these reasons, to increase the applicability of the method to solve the sought problem, we instead consider the case of a passive tether that does not require extra actuation. We remark that the results found in Chap. 4 are still valid. The only difference is that the link length is now not controllable but remains constant.

One of the main contributions of our work is the definition of some general conditions to perform a robust takeoff and landing. We then provide a careful analysis and a comparison of the contact-free flight and passive-tethered methods, based on these conditions. This study shows that, when an anchoring spot is available, the tether solution is highly preferable with respect to the contact-free flight one since it is the only one that allows to land on any sloped surface, and with good repeatability and robustness to tracking inaccuracies. Focusing on the passive-tether solution, in order to execute the maneuver respecting the inputs limits and to increase the robustness and safety of the maneuver, we also design a planner to compute an optimal reference trajectory. The latter is then followed by the hierarchical controller Γ^b_{HC} for the output y^b (see Sect. 4.5). We chose this controller rather then one based on dynamic feedback linearization, Γ^b_{DFL}, because highly dynamic trajectory are not required for the task (to increase the safety of the maneuver). The global method is finally tested through exhaustive real experiments in which a quadrotor is able to perform the landing and takeoff on/from a sloped surface tilted by an angle up to 60°. Part of the following results have been published in [5, 6].

6.2 Modeling

The unidirectional-thrust aerial vehicle is modeled as in Sect. 4.3 (link attached to a fixed platform and with a fixed length, i.e., no link actuator) and its states and control inputs are described by the same variables. We assume the vehicle equipped with at least three landers whose ending parts form the *landers plane* $\mathcal{P}_{RL}$. As in the most common case in reality, we assume $\mathbf{z}_R$ perpendicular to $\mathcal{P}_{RL}$.[1] Then we define $\mathbf{p}_{RL} \in \mathbb{R}^3$ as the projection of $\mathbf{p}_R$ on $\mathcal{P}_{RL}$ and $h_R = \|\mathbf{p}_R - \mathbf{p}_{RL}\|$ as the distance between $\mathbf{p}_R$ and $\mathcal{P}_{RL}$.

We assume that the *landing/takeoff (LTO) surface* is planar in the neighborhood of the desired landing point and it is defined by $\mathcal{P}_S := \{\mathbf{p} = [x \;\; y \;\; z]^\top \in \mathbb{R}^3 \mid ax + by + cz + d = 0\}$ where $a, b, c, d \in \mathbb{R}$ are the parameters of the plane. In particular, $\mathbf{n}_S = (1/\sqrt{a^2 + b^2 + c^2})[a \;\; b \;\; c]^\top$ are the coordinates in $\mathcal{F}_W$ of the unit vector normal to $\mathcal{P}_S$. Then we define a frame $\mathcal{F}_S$ that is rigidly attached to

[1]The equal interesting but unusual case of an arbitrary $\mathcal{P}_{RL}$ is left as future work.

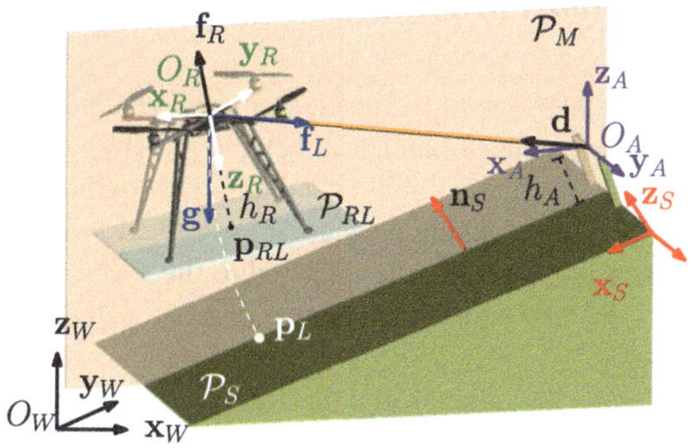

Fig. 6.1 Representation of the system and its main variables. © 2020 IEEE. Reprinted, with permission, from [5]

$\mathcal{P}_S$, whose axes are $\{\mathbf{x}_S, \mathbf{y}_S, \mathbf{z}_S\}$. If $\mathbf{n}_S = \mathbf{z}_W$, i.e., $\mathcal{P}_S$ is horizontal, then we set $\{\mathbf{x}_S, \mathbf{y}_S, \mathbf{z}_S\} = \{\mathbf{x}_W, \mathbf{y}_W, \mathbf{z}_W\}$. In the others (more interesting) cases, i.e., when $\mathcal{P}_S$ is locally inclined, the axes of $\mathcal{F}_S$ are set as: $\mathbf{z}_S = \mathbf{n}_S$, $\mathbf{y}_S = (\mathbf{z}_W \times \mathbf{z}_S)/\left\|\mathbf{z}_W \times \mathbf{z}_S\right\|$ and $\mathbf{x}_S = (\mathbf{y}_S \times \mathbf{z}_S)/\left\|\mathbf{y}_S \times \mathbf{z}_S\right\|$. The origin of $\mathcal{F}_S$, O_S, is taken as any arbitrary position on $\mathcal{P}_S$. Figure 6.1 gives a schematic representation of the whole system.

6.2.1 *Model in Free (Non-tethered) Flight*

Recalling the modeling of a unidirectional-thrust aerial vehicle in contact-free flight done in Sect. 3.2, its configuration is described by $\mathbf{p}_R$ and $\mathbf{R}_R$ and its dynamic is given by:

$$m_R\ddot{\mathbf{p}}_R = -m_R g \mathbf{z}_W - f_R \mathbf{z}_R \tag{6.1}$$

$$\mathbf{J}_R\dot{\boldsymbol{\omega}}_R = \mathbf{J}_R\boldsymbol{\omega}_R \times \boldsymbol{\omega}_R + \boldsymbol{\tau}_R. \tag{6.2}$$

Model (6.1) holds as long as the aerial vehicle is not in contact with the surface. In this last case, i.e., $\mathcal{P}_{RL} \equiv \mathcal{P}_S$, (6.1) has to be extended taking into account the reaction force of the surface, denoted by $f_N \in \mathbb{R}$, and the static friction force, denoted by $\mathbf{f}_S \in \mathbb{R}^3$, thus obtaining:

$$m_R\ddot{\mathbf{p}}_R = -m_R g \mathbf{z}_W - f_R \mathbf{z}_R + f_N \mathbf{n}_S + \mathbf{f}_S, \tag{6.3}$$

where $f_N \geq \underline{f_N}$, $\mathbf{z}_S^\top \mathbf{f}_S = 0$ and $\|\mathbf{f}_S\| \leq \overline{f_S}$. For a standard surface $\underline{f_N} = 0$ and $\overline{f_S} = \mu f_N$ where $\mu \in \mathbb{R}_{\geq 0}$ is the characteristic friction coefficient of the contact between $\mathcal{P}_{RL}$ and $\mathcal{P}_S$. If $\mathcal{P}_{RL}$ and $\mathcal{P}_S$ are equipped with an adhesive membrane (e.g., a Velcro or a gecko inspired material) then $\underline{f_N} \in \mathbb{R}_{\leq 0}$ is the maximum negative reaction force. In these cases both $\underline{f_N}$ and $\overline{f_S}$ depend on the adhesive membrane.

6.2.2 Model in Tethered Flight

Let us consider one of the particular tethered aerial cases considered in Sect. 4.3. In particular we consider an aerial vehicle tethered to a fixed point through a constant-length link, such as a cable or a chain. One end of the link is attached to the aerial vehicle at O_R through a passive 3D spherical joint and the other end is attached to an *anchor point* O_A rigidly attached to the surface. The position of O_A is described by $\mathbf{p}_A \in \mathbb{R}^3$ in $\mathcal{F}_W$ and its distance from $\mathcal{P}_S$ is given by $h_A = \mathbf{z}_S^\top(\mathbf{p}_A - \mathbf{p}_L) \in \mathbb{R}_{\geq 0}$, $\forall\, \mathbf{p}_L \in \mathcal{P}_S$.

When the link is slack and the aerial vehicle is not in contact with the LTO surface the dynamic model of the system is given by (6.1)–(6.2).

On the other hand, when the link is taut, the system model is the one presented in Sect. 4.3 when the link length is constant. We recall that $\mathbf{p}_R \in \mathcal{S}_l(\mathbf{p}_A) = \{\mathbf{p} \in \mathbb{R}^3 \mid \mathbf{p} = \mathbf{p}_A + l\mathbf{d},\ \forall \mathbf{d} \in \mathrm{S}^2\}$, where $\mathcal{S}_l(\mathbf{p}_A)$ is a sphere of radius l centered on $\mathbf{p}_A$, and $\mathbf{d}$ is the unit vector that represents the attitude of the link expressed in $\mathcal{F}_W$.

We introduce the frame $\mathcal{F}_A = \{O_A, \mathbf{x}_A, \mathbf{y}_A, \mathbf{z}_A\}$ defined as $\mathbf{z}_A = \mathbf{z}_W$, $\mathbf{y}_A = \mathbf{y}_S$ and $\mathbf{x}_A = \mathbf{y}_A \times \mathbf{z}_A / \|\mathbf{y}_A \times \mathbf{z}_A\|$. Recalling the modeling of Sect. 4.3 and assuming $\mathcal{F}_C = \mathcal{F}_A$, we have that the dynamics of the system is equal to (4.3) and (4.11) for the rotational and translational part, respectively. We recall that the model can be easily derived from (4.10) considering only the first three row, $\dot{l} = \ddot{l} = 0$ and replacing the notation C with the notation A where appropriate. For the reader convenience we report here the main equations with the proper notation that will be useful also in the following.

The vector $\mathbf{d}^A$ denotes the expression of $\mathbf{d}$ in $\mathcal{F}_A$. It is parametrized by the *elevation angle*, $\varphi \in [0, 2\pi]$, and the *azimuth angle*, $\delta \in [-\frac{\pi}{2}, \frac{\pi}{2}]$, such that $\mathbf{d}^A = [\cos\delta\cos\varphi \quad -\sin\delta \quad \cos\delta\sin\varphi]^\top$. Since the link is attached to O_R, the rotational dynamics of the vehicle is independent of the translational one and it is equal to (6.2). We retrieve the dynamics of[2] $\mathbf{q} = [\varphi\ \delta]^\top$ with the Newton–Euler method applying the balance of forces at O_R:

$$m_R\ddot{\mathbf{p}}_R = -m_R g\mathbf{z}_W - f_R\mathbf{z}_R - f_L\mathbf{d}, \tag{6.4}$$

where $\ddot{\mathbf{p}}_R$ is obtained differentiating twice $\mathbf{p}_R = \mathbf{p}_A + l\mathbf{R}_A\mathbf{d}^A$:

$$\ddot{\mathbf{p}}_R = \mathbf{R}_A\left(\dot{\mathbf{J}}_q\dot{\mathbf{q}} + \mathbf{J}_q\ddot{\mathbf{q}}\right), \quad \mathbf{J}_q = \begin{bmatrix} -l\cos\delta\sin\varphi & -l\cos\varphi\sin\delta \\ 0 & -l\cos\delta \\ l\cos\delta\cos\varphi & -l\sin\delta\sin\varphi \end{bmatrix},$$

where $\mathbf{R}_A \in \mathbb{R}^{3\times 3}$ is the rotation matrix from $\mathcal{F}_A$ to $\mathcal{F}_W$. Equations (6.2) and (6.4) fully describe the dynamics of the system when the link is taut.

[2] In this chapter, since the length of the link is constant, l is not a generalized variable but becomes a parameter of the system.

Similarly to the non-tethered case, when the robot is tethered and in contact with the surface, the model (6.4) is extended taking into account the reaction and friction forces, $f_N \in \mathbb{R}$ and $f_S \in \mathbb{R}$, respectively:

$$m_R \ddot{\mathbf{p}}_R = -m_R g \mathbf{z}_W - f_R \mathbf{z}_R - f_L \mathbf{d} + f_N \mathbf{n}_S + \mathbf{f}_S. \tag{6.5}$$

6.3 Conditions for Robust Landing and Takeoff

In the following we define and analyze the problem of landing on $\mathcal{P}_S$ at a *desired landing position* $\mathbf{p}_L^\star \in \mathcal{P}_S$. Analogous conditions can be drawn for the takeoff problem, which are omitted here for brevity. Denoting with $t_L \in \mathbb{R}_{>0}$ the *landing time*, a correct and robust landing is such if the following conditions are satisfied:

(1) $\mathbf{p}_{RL}$ converges to $\mathbf{p}_L^\star$, i.e., $\mathbf{p}_{RL}(t_L) = \mathbf{p}_L^\star \in \mathcal{P}_S$;
(2) the robot orientation has to be such that $\mathcal{P}_{RL}$ and $\mathcal{P}_S$ are parallel, i.e., $\mathbf{z}_R(t_L) = \mathbf{z}_R^\star = -\mathbf{z}_S$, in order to have the robot perfectly in contact with the surface;
(3) the vehicle has to reach this configuration with almost zero kinetic energy in order to avoid hard impacts, i.e., at time $t_L{}^-$, immediately before of touching the surface, it has to be that $\dot{\mathbf{p}}_R(t_L{}^-) = \mathbf{0}$ and $\boldsymbol{\omega}_R(t_L{}^-) = \mathbf{0}$;
(4) all the accelerations should be also zero at $t_L{}^-$, i.e., $\ddot{\mathbf{p}}_R(t_L{}^-) = \mathbf{0}$ and $\dot{\boldsymbol{\omega}}_R(t_L{}^-) = \mathbf{0}$, thus obtaining a smooth and gentle maneuver;

Definition (Inclined hovering) The system is said in *inclined hovering* if $\mathbf{z}_R \neq -\mathbf{z}_W$ and Conditions (3), and (4) coexist. □

(5) after the conclusion of the landing maneuver, at time $t_L{}^+$, when the robot is in contact with the surface, $\mathbf{p}_L^\star$ has to be a stable position, i.e., zero velocity and acceleration. This condition prevents the robot to fly away from the surface or to slide down on it when the motors are switched off after the landing maneuver.

Remark At time $t_L{}^-$ the robot is not yet in contact with the surface and the flying model has to be used to describe the system (Eqs. (6.1), (6.2) or (6.4), (6.2)). On the contrary, at time $t_L{}^+$ the vehicle is in contact with the surface thus Eqs. (6.3) or (6.5) have to be used. □

Notice that the Condition (4), although not strictly necessary, lets the vehicle approach the surface in a static equilibrium condition, passing from flight to contact very smoothly and in a more robust way with respect to model uncertainties.

If, due to the characteristics of the system, Condition (4) is not attainable, the landing can still be done but when at time $t_L{}^+$ the vehicle touches the surface, one has to find the way (e.g., turning off the motors as quickly as possible and using a Velcro system) to immediately pass in a stable condition in order to remain in contact with the surface without flying away or sliding on it (Condition (5)). Nevertheless, this could be not possible for some surfaces without the use of a tether or a Velcro-like solution.

Remark For the takeoff, only Condition (5), that now is an initial condition, has to be fulfilled. □

6.4 Analysis and Comparison for Landing and Takeoff

In the following we analyze two different kind of approaches for the landing: the free-flying and the tethered maneuvers. For both cases we define the conditions to satisfy the landing objectives and illustrate the benefits of the tethered solution.

6.4.1 Contact-Free Flight Method

Replacing the conditions $\mathbf{z}_S = \mathbf{z}_W$ and $\ddot{\mathbf{p}}_R({t_L}^-) = \mathbf{0}$ in (6.1), it is clear that the only case in which Condition (4) holds is when $\mathcal{P}_S$ is horizontal. In all the other cases $\ddot{\mathbf{p}}_R({t_L}^-) \neq \mathbf{0}$, which means that the aerial vehicle cannot approach the surface in a fully stable condition.

For the Condition (5), imposing $\ddot{\mathbf{p}}_R({t_L}^+) = \mathbf{0}$ in (6.3) and projecting the two sides of (6.3) on $\mathcal{F}_S$, we obtain

$$f_N = m_R g \mathbf{z}_S^\top \mathbf{z}_W + f_R, \quad \mathbf{x}_S^\top \mathbf{f}_S = m_R g \mathbf{x}_S^\top \mathbf{z}_W, \quad \mathbf{y}_S^\top \mathbf{f}_S = 0. \tag{6.6}$$

The first two conditions of (6.6) let us determine which is the maximum thrust at time ${t_L}^+$ and the maximum slope to have Condition (5) fulfilled, i.e.:

$$f_R({t_L}^+) \leq m_R g \mathbf{z}_S^\top \mathbf{z}_W - \underline{f_N} \quad \text{and} \quad \mathbf{x}_S^\top \mathbf{z}_W \leq \overline{f_S}/(m_R g). \tag{6.7}$$

Thus, one can land on any point of $\mathcal{P}_S$ only if (6.7) holds, restricting the set of admissible slopes.

Assuming that the surface fulfills (6.7), we now investigate how to reach it, and in particular, how to achieve the first three conditions. In the less interesting case of a horizontal surface, one can simply follow a trajectory along $\mathbf{z}_W$ in hovering condition to reach $\mathbf{p}_L^\star$ with zero velocities and accelerations. In the more interesting case of a sloped surface, this is a very challenging problem due to the underactuation of the vehicle. From the theory it is well known that the system is differentially flat with respect to $\mathbf{p}_R$ and the rotation around $\mathbf{z}_R$ [4]. Therefore one can track any desired position trajectory, $\mathbf{p}_R^d(t)$, such that $\mathbf{p}_{RL}(t_L) = \mathbf{p}_L^\star$ and $\dot{\mathbf{p}}_R({t_L}^-) = \mathbf{0}$, but the orientation of the vehicle along the trajectory is exactly determined by $\mathbf{p}_R^d(t)$ and its derivatives. Thus it is not possible to control the attitude independently from the position trajectory. The classical method to overcome this issue is to use a state-to-state planner like, e.g., the ones presented in [3] slightly modified, that gives a particular position trajectory $\mathbf{p}_R^d(t)$ that satisfies Conditions (1), (2) and (3).

Remark Consider an aerial vehicle that has to land on a given surface $\mathcal{P}_S$, at any desired point $\mathbf{p}_L^\star \in \mathcal{P}_S$. If the landing has to be performed using a contact-free flight method then, in general:

- if $\mathcal{P}_S$ is non-horizontal then Conditions (1), (2), (3) can only be achieved by a very accurate tracking of a perfectly synchronized dynamic maneuver generated using a state-to-state kino-dynamic planner;
- Condition (5) is fulfilled iff $f_R(t_L{}^+)$ and $\mathcal{P}_S$ are such that the two conditions in (6.7) hold;
- Condition (4) is fulfilled iff $\mathcal{P}_S$ is horizontal; □

Assuming that the non-easy motion planning problem is solved, one could use different types of controllers, as the ones in [1, 7], to track the planned trajectory. Nevertheless, these methods lack in general of robustness since small tracking errors could lead, e.g., to miss the target or to crash on it if the velocity is not well tracked. Furthermore, a precise model and an accurate and high-rate state estimation are needed.

To partially solve those problems and the ones related to the sliding, a common practical solution is to use a Velcro, as in [2, 8], to help the perching. However these solutions are not feasible in a real environments. Velcro solution also does not permit to easily takeoff after the perching.

6.4.2 *Tethered Method*

In this section we show that the tethered method overcomes the limits of contact-free flight (in particular the impossibility to satisfy Condition (4) for sloped surfaces, which guaranties a safer landing maneuver) thanks to the inclined equilibria.

For the tethered method the landing position must belong to $\mathcal{S}_l(\mathbf{p}_A) \cap \mathcal{P}_S$. We then first investigate which are the points in this set that satisfy Condition (4). Consider a generic point $\mathbf{p}_L \in \mathcal{S}_l(\mathbf{p}_A) \cap \mathcal{P}_S$. From simple geometry we have

$$\mathbf{d} = (\mathbf{p}_L - \mathbf{p}_A + h_R \mathbf{z}_S)/l. \tag{6.8}$$

Since $\mathbf{p}_A$, l and h_R are given parameters, finding the $\mathbf{p}_L$ that satisfies Condition (4) is equivalent to find the $\mathbf{d}$ that satisfies the same condition. Projecting both sides of (6.8) on $\mathbf{z}_S^\top$ we obtain

$$\mathbf{z}_S^\top \mathbf{d} = \left(h_R + \mathbf{z}_S^\top(\mathbf{p}_L - \mathbf{p}_A)\right)/l = (h_R - h_A)/l := c. \tag{6.9}$$

Then, in order to fulfill Condition (4), let us project both sides of (6.4) on the plane $\{\mathbf{x}_S, \mathbf{y}_S\}$, and set $\ddot{\mathbf{p}}_R = \mathbf{0}$, thus obtaining

$$f_L \mathbf{P}_{xy}^S \mathbf{d} = -m_R g \mathbf{P}_{xy}^S \mathbf{z}_W, \tag{6.10}$$

where $\mathbf{P}^S_{xy} = [\mathbf{x}_S\ \mathbf{y}_S]^\top$. Equation (6.10) implies that $\mathbf{P}^S_{xy}\mathbf{d}$ is parallel to $\mathbf{P}^S_{xy}\mathbf{z}_W$. Since $f_L \geq 0$ and $m_R g > 0$, we obtain

$$(\mathbf{P}^S_{xy}\mathbf{d})/\left\|\mathbf{P}^S_{xy}\mathbf{d}\right\| = -(\mathbf{P}^S_{xy}\mathbf{z}_W)/\left\|\mathbf{P}^S_{xy}\mathbf{z}_W\right\| = [1\ \ 0]^\top =: \mathbf{z}_W^{S_{xy}}. \tag{6.11}$$

Notice that (6.11) requires[3] $\left\|\mathbf{P}^S_{xy}\mathbf{z}_W\right\| \neq 0$ and $\left\|\mathbf{P}^S_{xy}\mathbf{d}\right\| \neq 0$. The latter inequality implies also that $(h_R - h_A) \neq l$. From (6.9) and (6.11) and applying some simple geometry we obtain

$$\mathbf{d} = [\mathbf{x}_S\ \mathbf{y}_S\ \mathbf{z}_S]\left[\sqrt{1-c^2}\ \ 0\ \ c\right]^\top =: \tilde{\mathbf{d}}, \tag{6.12}$$

where $\tilde{\mathbf{d}}$ is defined as the (unique) $\mathbf{d}$ for which Condition (4) is fulfilled. This proves that, given the parameters of the system, $\mathbf{p}_A$, l and h_R, it exists a (unique) $\tilde{\mathbf{p}}_L = \mathbf{p}_A + l\tilde{\mathbf{d}} - h_R\mathbf{z}_S$, for which Condition (4) is respected.

Remark The use of a tether creates the conditions to approach or depart from a sloped surface in a stable equilibria condition (*inclined hovering*), i.e., in a more robust and safer way. In fact, using the tether it exists a landing position in which one can land in *inclined hovering* for *any* sloped surface (in contact-free flight this position exists *only* for horizontal surfaces). Moreover, given any desired landing position $\mathbf{p}_L^\star \in \mathcal{P}_S$, one can always fulfill Condition (4) setting $h_A \neq h_R - l$ and

$$\mathbf{p}_A = \mathbf{p}_L^\star + h_R\mathbf{z}_S - l\tilde{\mathbf{d}} := \tilde{\mathbf{p}}_A. \tag{6.13}$$

□

6.4.2.1 Compliance with Condition (5)

If $(\mathbf{x}_S^\top\mathbf{z}_W)(\mathbf{x}_S^\top\mathbf{d}) < 0$, i.e., if the landing spot is below the projection of $\mathbf{p}_A$ on $\mathcal{P}_S$, then a solution of (6.5) for $\ddot{\mathbf{p}}_R({t_L}^+) = \mathbf{0}$ is

$$\begin{aligned} f_L &= -\frac{m_R g \mathbf{x}_S^\top\mathbf{z}_W}{\mathbf{x}_S^\top\mathbf{d}}, \quad f_N = m_R g \mathbf{z}_S^\top\mathbf{z}_W - f_R + f_L\mathbf{z}_S^\top\mathbf{d}, \\ \mathbf{y}_S^\top\mathbf{f}_S &= f_L\mathbf{y}_S^\top\mathbf{d}, \qquad \mathbf{x}_S^\top\mathbf{f}_S = 0. \end{aligned} \tag{6.14}$$

In this case the tension is always positive and, from the conditions on f_N and $\mathbf{f}_S$ in (6.14), we can determine which is the maximum thrust intensity at time ${t_L}^+$ and the maximum slope of the surface to respect the Condition (5), i.e.,

[3]When $\left\|\mathbf{P}^S_{xy}\mathbf{z}_W\right\| = 0$ the surface is horizontal and any $\mathbf{d}$ such that (6.9) holds, satisfies Condition (4). In this condition, one can still land with the tethered configuration keeping $f_L = 0$ and using the same method for contact-free flight.

Table 6.1 Characteristics of contact-free flight and tethered configuration for the landing problem. Analogous conditions hold for the take-off. © 2020 IEEE. Reprinted, with permission, from [5]

Method	Contact-free flight		Tethered flight	
Fulfillment of Condition	All	All but Condition (4) ($\ddot{\mathbf{p}}_R(t_L{}^-) \neq \mathbf{0}$)	All	All but Condition (4) ($\ddot{\mathbf{p}}_R(t_L{}^-) \neq \mathbf{0}$)
Surf. orientations	$\mathbf{z}_S = \mathbf{z}_W$	$\mathbf{x}_S^\top \mathbf{z}_W \leq \overline{f_S}/(m_R g)$	any	$\lvert -m_R g \mathbf{y}_S^\top \mathbf{d}(\mathbf{x}_S^\top \mathbf{z}_W \mathbf{x}_S^\top \mathbf{d}) \rvert \leq \overline{f_S}$
Anchor positions	–	–	any	any
Landing positions	any	any	$\mathbf{p}_L = \mathbf{p}_A + l\tilde{\mathbf{d}} - h_R \mathbf{z}_S$	$\mathbf{p}_L \in \mathcal{S}_l(\mathbf{p}_A) \cap \mathcal{P}_S$
Max. f_R at $t_L{}^+$	$f_R \leq m_R g - \underline{f_N}$	$f_R \leq m_R g \mathbf{z}_S^\top \mathbf{z}_W - \underline{f_N}$	$f_R(t_L{}^+) \leq \overline{f_R}$	$f_R(t_L{}^+) \leq \overline{f_R}$
	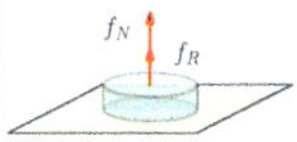	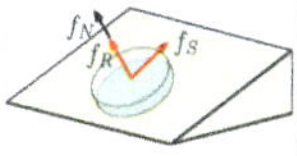	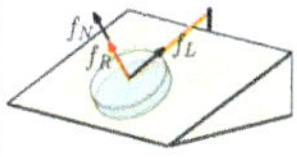	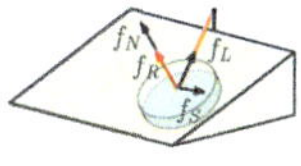
Pros	Simple system		Possibility to perform the maneuver reaching a stable equilibria condition; a planner is not required; robustness to model uncertainties and tracking errors	
Conditions	Not feasible for every slope; it requires: a planner, high tracking accuracy, precise state estimation and knowledge of the model (very low robustness)		Need of a method to pass from contact-free flight condition to tethered one	

$$f_R(t_L{}^+) \leq m_R g \mathbf{z}_S^\top \mathbf{z}_W \left(1 - (\mathbf{z}_S^\top \mathbf{d}/\mathbf{x}_S^\top \mathbf{d})\right) - \underline{f_N} =: \overline{f_R} \tag{6.15}$$

$$\left| -m_R g \mathbf{y}_S^\top \mathbf{d}(\mathbf{x}_S^\top \mathbf{z}_W/\mathbf{x}_S^\top \mathbf{d}) \right| \leq \overline{f_S}. \tag{6.16}$$

If $\mathbf{d} = \tilde{\mathbf{d}}$ then the Condition (6.16) holds for any surface. In the opposite case of $(\mathbf{x}_S^\top \mathbf{z}_W)(\mathbf{x}_S^\top \mathbf{d}) \geq 0$, i.e., when the landing spot is above the projection of $\mathbf{p}_A$ on $\mathcal{P}_S$, we have that $f_L = 0$ and the conditions in (6.7) have to be respected.

Table 6.1 summarizes all the previous results. To accomplish Conditions (1), (2) and (3) the controllers Γ^b_{IIC} or Γ^b_{DFL} presented in Sects. 4.5.2 and 4.6.2, respectively, can be used. Although not needed, to further improve the robustness and the reliability of the maneuver, we designed a motion planning method presented in the following Sect. 6.5 to optimize the motion during the landing and takeoff maneuvers. Notice that with the tethered method we can achieve all the landing conditions for any surface and any desired landing position by properly choosing the anchor point.

6.5 Optimal Trajectory Planning

Given the tracking controllers of Sects. 4.5.2 and 4.6.2 for the output $\mathbf{y}^b$, we need to design a feasible desired trajectory $\mathbf{y}^{bd}(t)$ that fulfills the objectives of Sect. 6.3 to successfully perform the landing. From now on we focus on the landing problem, since the trajectory for the takeoff can computed with the same method.

We assume that in a preliminary phase the vehicle has been tethered to the anchor point $\mathbf{p}_A$ such that $\mathbf{p}_L^\star = \tilde{\mathbf{p}}_L$, and the system has been steered to the state $\mathbf{x}_0$ for which the link results taut. Then, the initial and final value of the trajectory, $\mathbf{y}^{bd}(t)$, has to be such that $\mathbf{x}(t_0) = \mathbf{x}_0$ and $\mathbf{x}(t_L) = \mathbf{x}^\star$, where $\mathbf{x}^\star$ corresponds to the Conditions (1), (2), (3) and (4). In the following we define the final desired output value and an optimal planner to design a feasible and optimal trajectory that fulfills all the objectives of Sect. 6.3, and respects the input limits.

6.5.1 Final Desired Output

Since $\mathbf{p}_{RL}$ and $\mathbf{z}_R$ are independent from ψ, then $\psi^\star$ can be chosen arbitrarily. Given a desired landing position $\mathbf{p}_L^\star \in \mathcal{S}_l(\mathbf{p}_A) \cap \mathcal{P}_S$, one can compute the corresponding desired link attitude $\mathbf{d}^\star$ from (6.8). Finally, from the parametrization of $\mathbf{d}$ and (4.26) we can complete the remaining entries of the desired output $\mathbf{y}^{b\star}$:

$$
\begin{aligned}
\varphi^\star &= \operatorname{atan2}\left(\mathbf{z}_A^\top \mathbf{d}^\star, \mathbf{x}_A^\top \mathbf{d}^\star\right) \\
\delta^\star &= \operatorname{atan2}\left(\mathbf{y}_A^\top \mathbf{d}^\star, \sqrt{(\mathbf{z}_A^\top \mathbf{d}^\star)^2 + (\mathbf{x}_A^\top \mathbf{d}^\star)^2}\right) \\
\vartheta_A^\star &= \operatorname{atan2}\left(\alpha_1^\star, \alpha_3^\star\right),
\end{aligned}
$$

where $\alpha^\star = \mathbf{P}_L^\star \mathbf{z}_S$ and $\mathbf{P}_L^\star$ is computed as in Sect. 4.4.2 from $\mathbf{d}^\star$.

Notice that the equality $\mathbf{y}^b(t_L) = \mathbf{y}^{b\star} = [\varphi^\star \ \delta^\star \ \vartheta_A^\star \ \psi^\star]^\top$, does not necessarily imply that $\mathbf{z}_R(t_L) = \mathbf{z}_R^\star$. Indeed, controlling ϑ_A we control only the direction of the projection of $\mathbf{z}_R$ on the plane $\mathcal{P}_L$. Whereas, the remaining component $\mathbf{y}_L^\top \mathbf{z}_R$ is not directly controlled but, for the flatness, it is given by the particular trajectory $\mathbf{y}^b(t)$ and its derivatives. A possible solution consists on planning a proper trajectory $\mathbf{y}^b(t)$ such that $\mathbf{y}^b(t_L) = \mathbf{y}^{b\star}$ and $\mathbf{y}_L^\top \mathbf{z}_R(t_L) = \mathbf{y}_L{}^\top \mathbf{z}_R^\star = 0$. Though, this technique based on motion planning shows the same drawbacks saw in Sect. 6.4 for the contact-free flight method.

However, for the case of interest when all the objectives of Sect. 6.3 are fulfilled and in particular Condition (4), a planner is not necessary. In this case, from Eq. (6.12) and the parametrization of $\mathbf{d}$, it is easy to see that $\mathbf{d}^\star = \tilde{\mathbf{d}}$ implies $\delta^\star = 0$. Finally, thanks to the flatness, we can demonstrate that if δ is stabilized to zero, i.e., $[\delta, \dot{\delta}, \ddot{\delta}] = \mathbf{0}$, then $\mathbf{y}_L^\top \mathbf{z}_R$ is stabilized to the desired value $\mathbf{y}_L^\top \mathbf{z}_R^\star = 0$. Let us project (6.4) on $\mathbf{y}_L$ considering $[\delta, \dot{\delta}, \ddot{\delta}] = \mathbf{0}$. We obtain:

$$-f_R\mathbf{y}_L^\top\mathbf{z}_R = \mathbf{y}_L^\top\mathbf{R}_A\left[lm_R\left(\begin{bmatrix}-c_\varphi\\0\\s_\varphi\end{bmatrix}\dot{\varphi}-\begin{bmatrix}s_\varphi\\0\\c_\varphi\end{bmatrix}\ddot{\varphi}\right)+f_L\begin{bmatrix}c_\varphi\\0\\s_\varphi\end{bmatrix}\right]+m_Rg\mathbf{y}_L^\top\mathbf{z}_W. \quad (6.17)$$

Noticing that $\mathbf{y}_L^\top\mathbf{R}_A = \mathbf{e}_2^\top$ and $\mathbf{y}_L^\top\mathbf{z}_W = 0$, it is clear that $\mathbf{y}_L^\top\mathbf{z}_R = 0$.

This proves that, if $\mathbf{p}_A$ is chosen such that $\mathbf{p}_L^\star = \tilde{\mathbf{p}}_L$, then steering $\mathbf{y}^b$ to $\mathbf{y}^{b\star}$ is sufficient to steer $\mathbf{p}_{RL}$ and $\mathbf{z}_R$ to $\mathbf{p}_L^\star$ and $\mathbf{z}_R^\star$, respectively. In principle, we could generate a simple sufficiently smooth trajectory (like a spline) $\mathbf{y}^{bd}(t)$ from the initial output value to $\mathbf{y}^{b\star}$ that fulfill all the objectives of Sect. 6.3, and then track it with one of the controllers presented in Sects. 4.5.2 and 4.6.2, without the use of a planner. This makes the method more robust to tracking errors since they can be recovered by the controller avoiding the failure of the maneuver.

However, although not necessary, we still propose an optimal planner to more intuitively generate a desired trajectory that fulfills the conditions in Sect. 6.3, respecting the dynamics of the system, its input limits and other additional features in order to obtain an even more safe and reliable landing maneuver.

6.5.2 *Optimal Planner*

The computation of an optimal desired trajectory that fulfills the conditions of Sect. 6.3 can be formulated as an optimal control problem. Given the tethered system, we face a challenging nonlinear optimal control problem in a five dimensional configuration space. Even for a numerical solver it could be not easy to find a solution of this problem and its computation could require a lot of time. A common technique consists of simplifying the model of the system to make the problem solvable in a reasonable amount of time.

According to this method we assume[4] $\delta(t) = 0$ for all $t \in [t_0, t_L]$. In practice this fact limits the motion of the vehicle on the 2D vertical plane $\mathcal{P}_M = \{\mathbf{x}_A, \mathbf{z}_A\}$, reducing the configuration space to two dimensions. This is not a problem since we showed that if δ converges to zero then also $\mathbf{y}_L^\top\mathbf{z}_R$ converges to $\mathbf{y}_L{}^\top\mathbf{z}_R^\star = 0$. For simplicity we also assume a constant rotation around $\mathbf{z}_R$, i.e., $\psi(t) = \psi^\star$ for all $t \in [t_0, t_L]$. This means that the system, while moving on $\mathcal{P}_M$, is fully described only by φ and ϑ_A, equivalently to the reduced model presented in Sect. 4.3 (indeed, $\vartheta_A = \theta$). We define $\mathbf{x}_M = [x_{M1}\ x_{M2}\ x_{M3}\ x_{M4}]^\top = [\varphi\ \vartheta_A\ \dot{\varphi}\ \dot{\vartheta}_A]^\top$ and $\mathbf{u}_M = [f_R\ \tau_{Ry}^A]^\top$ the state and input of the 2D system, respectively. In particular $\tau_{Ry}^A \in \mathbb{R}$ is the torque applied by the robot along the axis $\mathbf{y}_A$, i.e., $\tau_{Ry}^A = \mathbf{y}_A^\top\mathbf{R}_R\boldsymbol{\tau}_R$. Considering δ, $\mathbf{y}_A^\top$ and their derivatives to zero, the dynamics can be derived as done in Sect. 4.8

$$\begin{bmatrix}\ddot{\varphi}\\ \ddot{\vartheta}_A\end{bmatrix} = \begin{bmatrix}a_1\cos\varphi + a_2\cos(\varphi+\vartheta_A)f_R\\ a_3\tau_{Ry}^A\end{bmatrix} =: \mathbf{f}_M(\mathbf{x}_M, \mathbf{u}_M) \quad (6.18)$$

where $a_1 = -g/l$, $a_2 = 1/(m_R l)$, $a_3 = 1/(\mathbf{y}_A^\top\mathbf{R}_R\mathbf{J}_R)$.

[4]This can be guaranteed by the controllers proposed in Sects. 4.5.2 and 4.6.2.

From now on we focus on the generation of trajectories for a quadrotor-like VTOL since this is the robot used for the real experiments[5] described in Sect. 6.6. As recalled in Sects. 3.2 and 3.3, this particular vehicle is equipped with four propellers, placed in a symmetric configuration with respect to the center of gravity, each one generating a thrust $f_i \in [\underline{f}_i, \overline{f}_i]$. All together the propellers generate the total thrust f_R and torque $\boldsymbol{\tau}_R$ applied to the vehicle according the relation $\mathbf{u} = [f_R \ \boldsymbol{\tau}_R]^\top = \Gamma \mathbf{u}_f$ where $\mathbf{u}_f = [f_1 \ f_2 \ f_3 \ f_4]^\top$ and $\Gamma \in \mathbb{R}^{4\times4}$ is a matrix that maps $\mathbf{u}_f$ into $\mathbf{u}$ (see Sect. 3.3.1). The matrix Γ depends on the parameters of the vehicle. We can then express $\mathbf{u}_M$ as function of $\mathbf{u}_f$ as

$$\mathbf{u}_M = \begin{bmatrix} 1 & \mathbf{0} \\ 0 & \mathbf{y}_A^\top \mathbf{R}_R \end{bmatrix} \Gamma \mathbf{u}_f = \Gamma_M \mathbf{u}_f. \tag{6.19}$$

Notice that Γ_M is constant since, given the constraint of moving on $\mathcal{P}_M$, the vehicle body rotates only around $\mathbf{y}_A$. Then, replacing (6.19) into (6.18) we can define $\mathbf{f}_{Mf}(\mathbf{x}_M, \mathbf{u}_f) = \mathbf{f}_M(\mathbf{x}_M, \Gamma_M \mathbf{u}_f)$.

It is well known that the propellers of a real quadrotor can not immediately actuate a commanded thrust and that the time response depends on the particular motors and propellers. In order to obtain a feasible and smoother trajectory for the system we decided to consider a double dynamic extension of the model assuming as new input the second derivative of the thrust, $\bar{\mathbf{u}}_f = [\ddot{f}_1 \ \ldots \ \ddot{f}_4]^\top$. The new extended state becomes $\bar{\mathbf{x}}_M = [\varphi \ \vartheta_A \ \dot{\varphi} \ \dot{\vartheta}_A \ \ddot{\varphi} \ \ddot{\vartheta}_A \ \varphi^{(3)} \ \vartheta_A^{(3)} \ \mathbf{u}_f^\top \ \dot{\mathbf{u}}_f^\top]^\top$. The dynamics becomes

$$\dot{\bar{\mathbf{x}}}_M = \begin{bmatrix} \mathbf{0} & \mathbf{I}_6 & \mathbf{0} & \mathbf{0} \\ \mathbf{0} & \mathbf{0} & \mathbf{0} & \mathbf{0} \\ \mathbf{0} & \mathbf{0} & \mathbf{I}_4 & \mathbf{0} \\ \mathbf{0} & \mathbf{0} & \mathbf{0} & \mathbf{0} \end{bmatrix} \bar{\mathbf{x}}_M + \begin{bmatrix} \mathbf{0} \\ \ddot{\mathbf{f}}_{Mf}(\bar{\mathbf{x}}_M, \bar{\mathbf{u}}_f) \\ \mathbf{0} \\ \bar{\mathbf{u}}_f \end{bmatrix} = \bar{\mathbf{f}}_{Mf}(\bar{\mathbf{x}}_M, \bar{\mathbf{u}}_f). \tag{6.20}$$

We highlight the fact that for a very reactive vehicle characterized, e.g., by a low mass and inertia, this dynamic extension could be avoided, since it would be able to actuate fast varying inputs. Nevertheless, this allows us to generate a C^3 trajectory required by Γ^b_{DFL}.

Given the system dynamics (6.20), we are ready to formalize our optimal control problem as

$$\begin{aligned} &\min_{\bar{\mathbf{x}}_M(t), \bar{\mathbf{u}}_f(t)} \quad J(\bar{\mathbf{x}}_M(t), \bar{\mathbf{u}}_f(t), t, t_L) \\ &\text{subject to, } \forall \ t \in [t_0, t_L] \\ &(a) \ \ \dot{\bar{\mathbf{x}}}_M = \bar{\mathbf{f}}_{Mf}(\bar{\mathbf{x}}_M(t), \bar{\mathbf{u}}_f(t)) \ \ (b) \ \ \bar{\mathbf{x}}_M(t_0) = \bar{\mathbf{x}}_{M0} \\ &(c) \ \ \underline{f}_i \le f_i(t) \le \overline{f}_i \qquad\qquad (d) \ \ f_L(t) > 0 \end{aligned}, \tag{6.21}$$

[5] The method can be easily modified according to any VTOL.

where $J : \bar{\mathbf{x}}_M, \bar{\mathbf{u}}_f \rightarrow \mathbb{R}$ is the cost function, (a) is the dynamics, (b) are the initial conditions, (c) are the input limits, and (d) prevents the link to become slack. In order to fulfill the objectives of Sect. 6.3, we define the cost function as:

$$J = \int_{t_0}^{t_L} (J_1 + J_2 + J_3 + J_4) d\tau + J_5$$

where

$$\begin{aligned}
J_1 &= k_\varphi(\varphi - \varphi^\star)^2 + k_{\vartheta_A}(\vartheta_A - \vartheta_A^\star)^2 + k_{\dot{\varphi}}\dot{\varphi}^2 + k_{\dot{\vartheta}_A}\dot{\vartheta}_A^2 + k_{\ddot{\varphi}}\ddot{\varphi}^2 + k_{\ddot{\vartheta}_A}\ddot{\vartheta}_A^2 \\
J_2 &= k_{\vartheta_A 2} h_{\vartheta_A}(\varphi)(\vartheta_A - \vartheta_A^\star)^2 \\
J_3 &= k_{\dot{\varphi}\dot{\vartheta}_A} h_{\dot{\varphi}\dot{\vartheta}_A}(\varphi)(\dot{\varphi}^2 + \dot{\vartheta}_A^2) \\
J_4 &= k_{\bar{\mathbf{u}}_f} \bar{\mathbf{u}}_f^2 \\
J_5 &= k_{t_L \vartheta_A} \dot{\vartheta}_A^2 + k_{t_L \varphi} \dot{\varphi}^2
\end{aligned}$$

and $k_\star \in \mathbb{R}_{\geq 0}$, $h_{\vartheta_A}(\varphi)$ and $h_{\dot{\varphi}\dot{\vartheta}_A}(\varphi)$ are functions that tend to 1 when φ is near $\varphi^\star$, and to zero otherwise. The cost terms 1, 3, 5 together, help to fulfill the conditions of Sect. 6.3, i.e., to steer the vehicle on the surface approaching it with zero velocities and accelerations. The cost term 2 enforces to approach $\mathcal{P}_S$ with the proper attitude, such that the landers touch the surface simultaneously. Finally, the cost term 4 avoids fast variations on the commanded thrust that otherwise could not be actuated. Modifying the gains of J one can adjust the trajectory to obtain different behaviors.

The solution of the optimal control problem, $\bar{\mathbf{x}}_M^d(t)$ for $t \in [t_0, t_L]$, is computed using the ACADO [9] numerical optimizer. Finally, $\bar{\mathbf{x}}_M^d(t)$ together with $\delta^d(t) = 0$ and $\psi^d(t) = \psi^\star$ give the desired output trajectory $\mathbf{y}^{bd}(t)$ to be tracked in order to perform the landing.

6.6 Experimental Landing and Takeoff

In this section we show the main results of the experiments that validate the efficacy of our proposed method for the problem of landing (and takeoff) on a sloped surface.

In particular, we consider the plausible scenario where a quadrotor-like vehicle has to deploy a smaller robot or a sensor on a sloped surface tilted by 50°, shown in Fig. 6.4. The robot, equipped with a cable ending with a hook, starts from a non-tethered configuration on the ground. Therefore it has to anchor the other end of the cable to the surface to then perform the landing in a tethered configuration. Once the robot has landed on the desired spot and deployed the robot/sensor, it can take-off from the surface again exploiting the tether. Finally it can go back to the initial position after having detached the cable from the surface. The hardware employed for the experiment is the one described in Sect. 5.1.2.

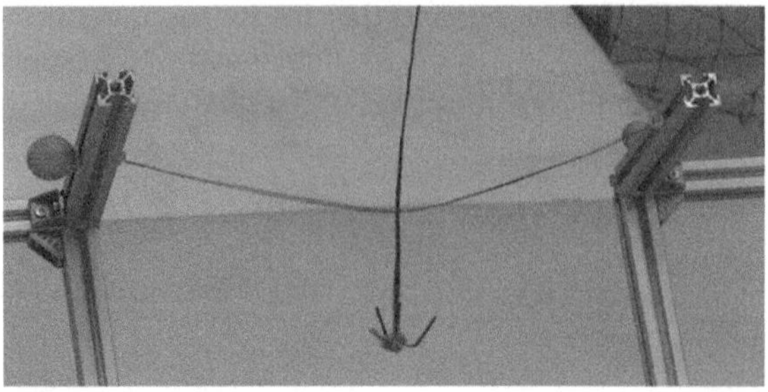

Fig. 6.2 Zoom of the hook and the anchoring mechanism

6.6.1 Anchoring Tools and Mechanisms

In order to pass from a contact-free flight configuration to a tethered one, a method to fix the end of the cable to the surface has to be found. The mechanism to do so strongly depends on the application scenario and in particular on the material of the slope. For example, in the context of the European project *Aeroarms*,[6] an aerial robot has to deploy a magnetic crawler or a sensor on industrial pipes that are often non-horizontal. In this context the landing surface is mainly a pipe made of iron/steel. Thus in this case, and whenever the surface is made of proper metal, a magnetic anchor can be used to enhance the physical connection between surface and the robot. In the case of a ground, snowed, or iced surface an harpoon-like mechanism might be envisaged.

In our experimental testbed we instead used a simpler solution based on a commercial fishing hook made of three tips, and an anchoring mechanism fixed to the surface made by a horizontal cable. In this way the robot can be tethered to the surface by sliding the vertical cable on the anchoring mechanism until the hook is anchored to the horizontal cable, as shown in Fig. 6.2. The hook can be detached from the anchoring mechanism doing the opposite operation.

6.6.2 Experimental Phases

Considering the previous experimental scenario and the goal, we divided the overall maneuver into several phases:

(a) approach to the anchor point with the hook,
(b) hooking of the anchoring system,
(c) stretching of the cable,
(d) tracking of the desired trajectory for tethered landing.

The phases from (a) to (c), described by the first row of images in Fig. 6.4, serve to pass from the initial contact-free flight configuration to the tethered one. Using a

[6]http://www.aeroarms-project.eu/.

standard contact-free flight position controller and following a straight-line trajectory, the robot is able to anchor the anchoring system attached to the surface with the hook (see Fig. 6.4b.2). The trajectory is planned such that the cable attached to the robot slides on the anchoring cable until the hook results attached to the last one.

Afterword, during phase (c), the cable is stretched following a simple radial trajectory whose ending point is slightly outside the reachable region limited by the cable length. The robot, trying to reach this ending position, as explained in [10], will apply an extra force to the cable that will make it taut. In particular, the farther the desired ending position, the larger the internal force on the link. Using the dynamics of the system, the estimated state, and the control inputs, the robot can estimate the tension on the link. This estimation is then used to understand when the cable results sufficiently taut. Once the tension exceeds a certain security threshold a supervisor switches from the contact-free flight controller to the tethered one. We recall that for this experiment we use the hierarchical controller Γ^{b}_{HC} presented in Sect. 4.5.2 whose validity has been experimentally demonstrated in Sect. 5.2. In fact, Γ^{b}_{HC} can guarantees sufficiently small tracking errors for the slow trajectories needed for the safety of the maneuvers. This additionally shows that the tethered solution does not require a very precise tracker.

Finally the planned landing trajectory is tracked. In order to compute the desired landing and takeoff trajectories using the planner presented in Sect. 6.5, the parameters of the landing surface, such as slope angle and anchoring point, must be known. To acquire those values we applied some markers on the surface to measure its pose with a motion capture system. However, thanks to the robustness of the method, those parameters does not have to be very precise.

Once the robot ends the landing maneuver the takeoff can start. The takeoff maneuver is very similar to the play-back of the previous phases. Indeed, following the previous trajectory in the opposite sense lets the hook be detached from the anchoring mechanism to then go back to the starting point in a contact-free flight configuration.

6.6.3 Controller Switch

During the switching between the controllers, the continuity of the control input has to be guaranteed in order to preserve the stability of the system and to avoid undesired vibrations and jerks on the cable. This is obtained by setting as desired output of the next controller, the value of the system output at the switching instant. This is possible because, thanks to the flatness, there is a bijective relation between state/input and output. Therefore, for a specific output, there exist a unique nominal input and state to obtain it. Assume that the system is in a certain state with a certain input, $\mathbf{x}_0$ and $\mathbf{u}_0$, respectively. Accordingly we have a particular output value $\mathbf{y}_0$. Asking the next controller to keep the output value $\mathbf{y}_0$ we will obtain the same input $\mathbf{u}_0$ and state $\mathbf{x}_0$, thus preserving the continuity of the control action and of the full state.

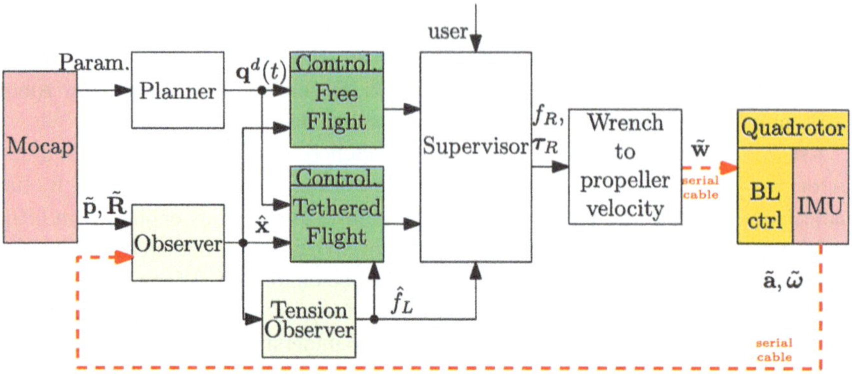

Fig. 6.3 Schematic representation of the software architecture. Pink blocks represent the sensors. Green blocks represent the controllers and light yellow blocks represent the observers. Starting from the left, $\tilde{\mathbf{p}}$ and $\tilde{\mathbf{R}}$ represent the measured robot position and orientation, respectively; $\hat{\mathbf{x}}$ and $\hat{f}_L$ represent the estimated state and link internal force, respectively; $\mathbf{q}^d(t)$ represents the desired output trajectory; f_R and $\boldsymbol{\tau}_R$ represents the input of the robot, i.e., thrust intensity and torque vector; $\tilde{\boldsymbol{w}}$ represents the desired spinning velocity of the propellers. Finally $\tilde{\mathbf{a}}$ and $\tilde{\boldsymbol{\omega}}$ represent the readings of the IMU, i.e., specific acceleration and angular velocity

6.6.4 *Software Architecture*

A schematic representation of the software architecture is represented in Fig. 6.3. The overall controllers and observers run on a ground PC. The desired spinning velocities of each propeller are sent at 500 Hz to the robot using a serial cable. The received velocity commands are then actuated by a controller (presented in [11]) running on the on-board ESC (Electronic Speed Control). The same serial communication is used to read at 1 KHz the IMU measurements that are then UKF-fused together with the motion capture system measurements (position and orientation of the quadrotor at 120 Hz) to obtain an estimation of the state of the vehicle. The latter is then used to close the control loop and to compute an estimation of the internal force along the link when it is taut.

The controller for the contact-free flight and tethered cases run in parallel and a supervisor, according to the state of the experiment, decides whose input has to be applied to the real system. The user input in the supervisor is needed to trigger situations of emergency.

6.6.5 *Offset Nonideality*

Another practical aspect that has to be considered is the nonzero offset between the cable attaching point and the vehicle center of mass. Indeed, the controller Γ^b_{HC} presented in Sect. 4.5.2 assumes that this offset is equal to zero. In this way the robot translational and rotational dynamics can be decoupled. However, this never happens in a practical case. Then, due to this non-zero offset, the internal force along the cable generates a torque on the vehicle that has to be carefully compensated. This is done computing the extra torque from the estimated tension and the estimated offset calculated with a mechanical analysis.

Finally we highlight the fact that the maximum tiling of the surface is bounded by the input limits. Indeed the more inclined is the slope, the less it is the thrust required to compensate the gravity close to the surface. Due to the impossibility of producing negative thrust for the single propeller, the almost zero total thrust implies a reduced control authority on the total input moment that may cause the instability of the attitude dynamics and of the whole system in general.

6.6.6 Experimental Results

In Figs. 6.4 and 6.5 the experimental results are shown. In this particular case the robot has to land and then takeoff on/from a planar surface that is tilted by 50°. Figure 6.4 shows the first half of the experiment, i.e., the landing, by a series of images. In particular the first row shows the anchoring procedure done in a contact-free flight condition. On the other hand, the second row shows the actual execution of the tethered landing. A video of the full experiment is available at [12].

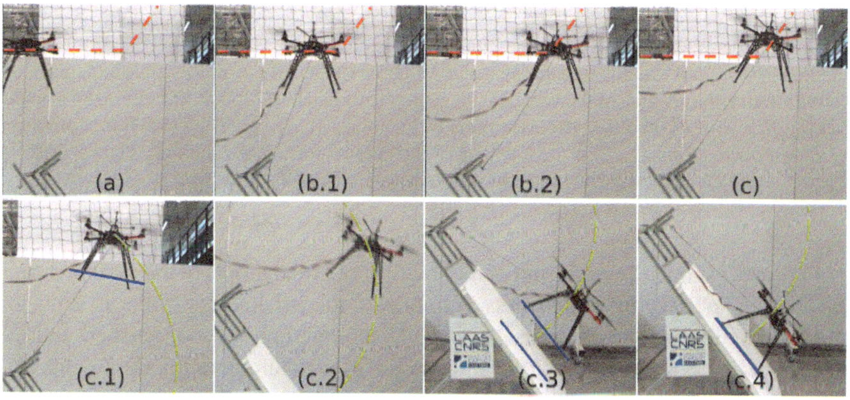

Fig. 6.4 Sequence of images of a real experiment with a sloped surface tilted by 50°. The first row of images represents the experimental part in which the quadrotor is in a contact-free flight condition. In this case a standard position controller is used to track the desired position trajectory marked with a dashed red line. The second row of images represents the experimental part in which the quadrotor is tethered to the surface. In this case the controller proposed in Sect. 4.5.2 is used to track the desired position and attitude trajectories marked with a dashed yellow line and a solid blue line, respectively

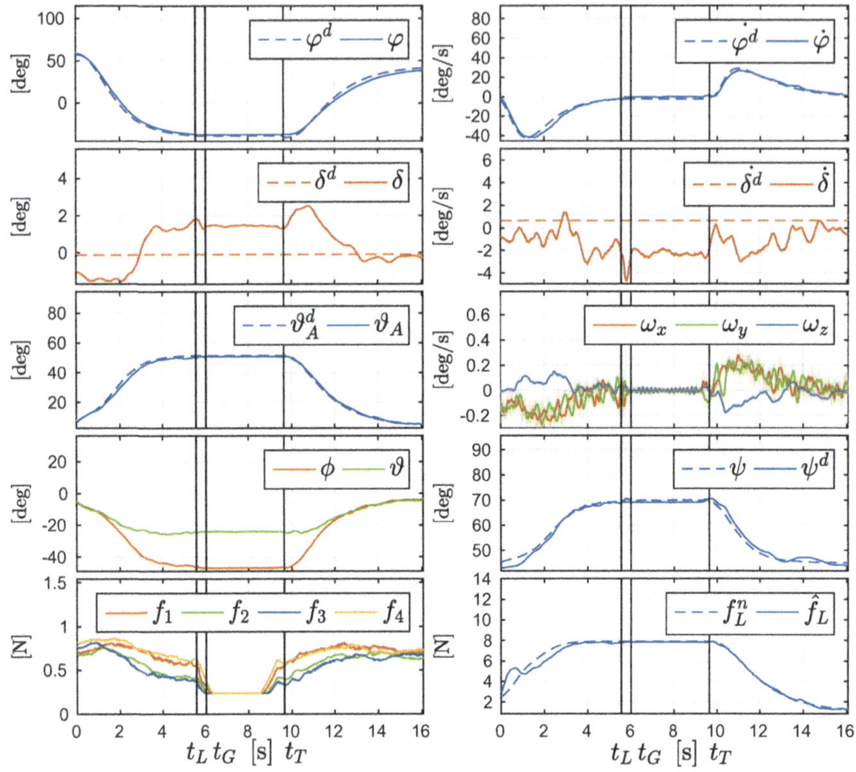

Fig. 6.5 Experimental results: plots of the state, outputs and inputs of the system during the tethered landing and takeoff. In particular φ and δ describe the attitude of the cable and, given the link constraint, the position of the vehicle with respect to the anchoring point. ϕ and ψ are the angles that together with ϑ_A describe the orientation of the robot. f_1, f_2, f_3, f_4 are the forces produced by each propeller. Finally, f_L is the intensity of the internal force along the link. The super-script d and n represent the desired and the nominal values of a variable, respectively

Figure 6.5 shows the evolution of the state, outputs and inputs of the system during the landing and takeoff maneuvers. At time zero the tethered controller is activated and the landing maneuver starts. At time t_L the landing is accomplished and the surface is reached. At time t_G the motors are stop to simulate the deploying of a robot/sensor. Finally, at time t_T the takeoff maneuver starts.

From those plots one can see that the desired trajectory is tracked precisely, with only some small errors due to calibration inaccuracy. Furthermore, notice that the intensity of the internal force is always positive. This shows that the cable is kept taut for the whole execution of the maneuvers. Despite the presence of tracking errors the landing and takeoff maneuver are accomplished successfully and in a very safe and gentle way. This shows the big advantage of using a tether that makes the execution

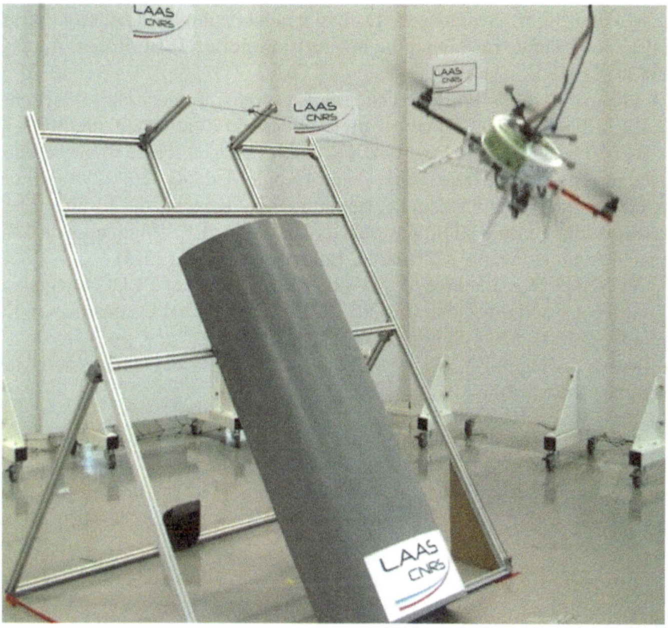

Fig. 6.6 Execution of a tethered landing on an inclined pipe tilted by 60°

on the task reliable and robust to tracking and modeling errors. Thanks to this we were able to perform landing and takeoff on/from an even non-flat surface (a pipe) tilted up to 60°. We provide an image from such experiment in Fig. 6.6.

References

1. Cabecinhas, D., Naldi, R., Marconi, L., Silvestre, C., Cunha, R.: Robust take-off for a quadrotor vehicle. IEEE Trans. Robot. **28**(3), 734–742 (2012)
2. Mellinger, D., Michael, N., Kumar, V.: Trajectory generation and control for precise aggressive maneuvers with quadrotors. Int. J. Robot. Res. **31**(5), 664–674 (2012)
3. Mueller, M.W., Hehn, M., D'Andrea, R.: A computationally efficient algorithm for state-to-state quadrocopter trajectory generation and feasibility verification. In: 2013 IEEE/RSJ International Conference on Intelligent Robots and Systems, pp. 3480–3486, Tokyo, Japan (2013)
4. Martin, P., Devasia, S., Paden, B.: A different look at output tracking: control of a VTOL aircraft. Automatica **32**(1), 101–107 (1996)
5. Tognon, M., Testa, A., Rossi, E., Franchi, A.: Takeoff and landing on slopes via inclined hovering with a tethered aerial robot. In: 2016 IEEE/RSJ International Conference on Intelligent Robots and Systems, pp. 1702–1707, Daejeon, South Korea (2016)
6. Tognon, M., Franchi, A.: Landing and take-off on/from sloped and non-planar surfaces with more than 50 degrees of inclination. In: 9th International Micro Air Vehicles Conference, pp. 97–102 (2017)

7. Faessler, M., Franchi, A., Scaramuzza, D.: Differential flatness of quadrotor dynamics subject to rotor drag for accurate tracking of high-speed trajectories. IEEE Robot. Autom. Lett. **3**(2), 620–626 (2018)
8. Mohta, K., Kumar, V., Daniilidis, K.: Vision-based control of a quadrotor for perching on lines. In: 2014 IEEE International Conference on Robotics and Automation, pp. 3130–3136 (2014)
9. Houska, B., Ferreau, H., Diehl, M.: ACADO Toolkit - an open source framework for automatic control and dynamic optimization. Opt. Control Appl. Methods **32**(3), 298–312 (2011)
10. Gioioso, G., Ryll, M., Prattichizzo, D., Bülthoff, H.H., Franchi, A.: Turning a near-hovering controlled quadrotor into a 3D force effector. In: 2014 IEEE International Conference on Robotics and Automation, pp. 6278–6284, Hong Kong, China (2014)
11. Franchi, A., Mallet, A.: Adaptive closed-loop speed control of BLDC motors with applications to multi-rotor aerial vehicles. In: 2017 IEEE International Conference on Robotics and Automation, pp. 5203–5208, Singapore (2017)
12. video 1: Takeoff and landing on slopes via inclined hovering with a tethered aerial robot. https://www.youtube.com/watch?v=01UYN289YXk (2016)

Chapter 7
Towards Multiple Tethered Aerial Vehicles

7.1 Modeling

In order to refer to the quantities of one component of the chain, we use the subscript $\bullet_i$ with $i = 1$ for the first link and $i = 2$ for the second. Similarly to Chap. 4 we assume: (i) negligible link masses and rotational inertias with respect to the ones of the vehicles, (ii) fixed link lengths $l_i \in \mathbb{R}_{>0}$ where $i \in \{1, 2\}$, and (iii) negligible deformations and elasticities.

We define $\varphi_i \in \mathbb{R}$ the elevation angle of the ith link. With $f_{Li} \in \mathbb{R}$ we denote the internal force that is exerted on the ith link. Also in this case the link is generic and both compressions and tensions are allowed. The first link is connected at one end to the CoM of the first vehicle, and the other end to a *fixed point*. The two ends of the second link are attached to the first and second vehicle center of masses, respectively. No rotational constraints are present in the connections, e.g., by using passive rotational joints. Finally, $m_{Ri} \in \mathbb{R}_{>0}$ and $J_{Ri} \in \mathbb{R}_{>0}$, with $i = 1, 2$, denote the mass and inertial, respectively, of the ith vehicle.

It is convenient to define the frames of the system in 3D, even if we consider a 2D problem, in order to, e.g., have a well defined angular velocity vector for the aerial vehicles. Thus we define a *world frame*, $\mathcal{F}_W$, described by the unit vector along its axes $\{\mathbf{x}_W, \mathbf{y}_W, \mathbf{z}_W\}$ and origin set on a fixed point O_W. Then, for every robot, we define a *body frame*, $\mathcal{F}_{Bi}$, rigidly attached to the ith vehicle, described by the unit vector along its axes $\{\mathbf{x}_{Bi}, \mathbf{y}_{Bi}, \mathbf{z}_{Bi}\}$ and origin O_{Bi} set on the vehicle CoM, represented in $\mathcal{F}_W$ by the coordinates $\mathbf{p}_{Ri} = \begin{bmatrix} x_{Bi} & y_{Bi} & z_{Bi} \end{bmatrix}^\top$, where $y_{Bi} = 0$. The axes $\mathbf{y}_W$, $\mathbf{y}_{B_1}$ and $\mathbf{y}_{B_2}$ are perpendicular to the vertical plane $\{\mathbf{x}_W, \mathbf{z}_W\}$ where motion occurs, as depicted in Fig. 7.1. The system evolves on this vertical plane on the effect of the four control inputs (two for each robot), i.e., the intensities $f_{Ri} \in \mathbb{R}$ and $\tau_{Ri} \in \mathbb{R}$ of the thrust force $-f_{Ri}\mathbf{z}_{Bi} \in \mathbb{R}^3$ and the torque $-\tau_{Ri}\mathbf{y}_{Bi} \in \mathbb{R}^3$, respectively, with $i = 1, 2$.

M. Tognon and A. Franchi, *Theory and Applications for Control of Aerial Robots in Physical Interaction Through Tethers*, Springer Tracts in Advanced Robotics 140,
https://doi.org/10.1007/978-3-030-48659-4_7

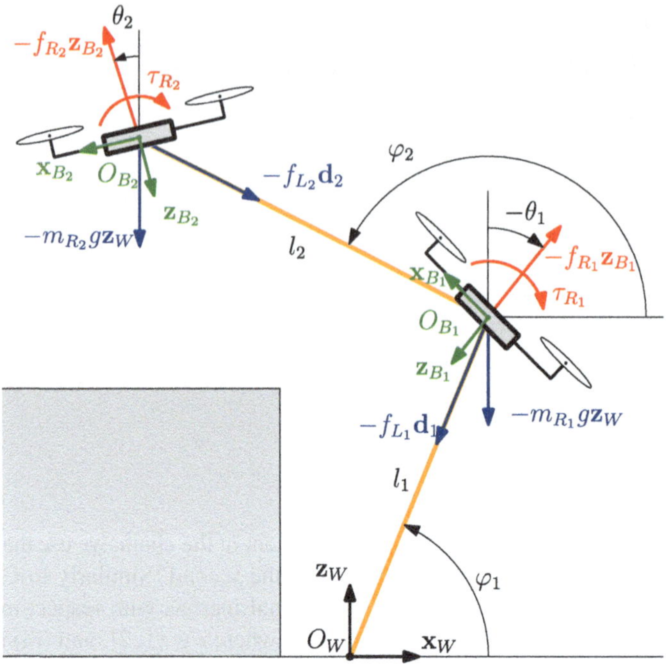

Fig. 7.1 Representation of the system and its main variables. The system is depicted in a scenario of example where the grey box represents a surface of manipulation for, e.g., a pick and place task.

Given the constraints, the system is completely described by the generalized coordinates $\mathbf{q} = [\varphi_1 \;\; \varphi_2 \;\; \theta_1 \;\; \theta_2]^\top = [\boldsymbol{\varphi}^\top \;\; \boldsymbol{\theta}^\top]^\top \in \mathbb{R}^4$, where φ_i and θ_i are the *elevation* of the ith link (defined before) and the *attitude* of the ith vehicle, respectively.

To derive the dynamic model of the system, as done in Sect. 4.3, we employ the Newton–Euler methods, because also in this case we are interest in controlling the internal force along the link and an analytical expression is thus needed. Since the rotational dynamics of the generic ith vehicle is decoupled by the translational one, we have that

$$\ddot{\boldsymbol{\theta}} = J^{-1}\boldsymbol{\tau}_R, \tag{7.1}$$

where $J = \mathrm{diag}(J_{R1}, J_{R2}) \in \mathbb{R}^{2\times 2}_{>0}$ and $\boldsymbol{\tau}_R = [\tau_{R1} \;\; \tau_{R2}]^\top \in \mathbb{R}^2$. Since we are considering the 2D problem, in the following, we will omit the lines full of zeros relative to the $\mathbf{y}_{B_1}$ and $\mathbf{y}_{B_2}$ axes. Balancing the forces acting on the vehicle CoMs we obtain

$$\underbrace{\begin{bmatrix} m_1\ddot{\mathbf{p}}_{B_1} \\ m_2\ddot{\mathbf{p}}_{B_2} \end{bmatrix}}_{\mathbf{a}} = -\underbrace{\begin{bmatrix} \mathbf{d}_1 f_{L1} - \mathbf{d}_2 f_{L2} \\ \mathbf{d}_2 f_{L2} \end{bmatrix}}_{\mathbf{a}_{\mathbf{f}_L}} - \underbrace{\begin{bmatrix} f_{R1}\mathbf{z}_{B_1} \\ f_{R2}\mathbf{z}_{B_2} \end{bmatrix}}_{\mathbf{a}_{\mathbf{f}_R}} - \underbrace{\begin{bmatrix} m_{R1}g\mathbf{z}_W \\ m_{R2}g\mathbf{z}_W \end{bmatrix}}_{\mathbf{a}_{\mathbf{g}}}, \tag{7.2}$$

where $\mathbf{d}_i = [\cos\varphi_i \;\; \sin\varphi_i]^\top$ and $\mathbf{d}_i^\perp = [-\sin\varphi_i \;\; \cos\varphi_i]^\top$ are unit vectors in the vertical plane parallel and perpendicular to the ith link, respectively. The accelerations of the vehicle CoMs expressed in $\mathcal{F}_W$ are

$$\begin{aligned}\ddot{\mathbf{p}}_{B_1} &= -l_1\mathbf{d}_1\dot{\varphi}_1^2 + l_1\mathbf{d}_1^\perp\ddot{\varphi}_1\\ \ddot{\mathbf{p}}_{B_2} &= \ddot{\mathbf{p}}_{B_1} - l_2\mathbf{d}_2\dot{\varphi}_2^2 + l_2\mathbf{d}_2^\perp\ddot{\varphi}_2.\end{aligned} \tag{7.3}$$

Using (7.3) and (7.2) we have that

$$\mathbf{a} = \underbrace{\begin{bmatrix} -m_{R1}l_1\mathbf{d}_1\dot{\varphi}_1^2 \\ -m_{R2}(l_1\mathbf{d}_1\dot{\varphi}_1^2 + l_2\mathbf{d}_2\dot{\varphi}_2^2)\end{bmatrix}}_{\mathbf{a}_{\dot{\varphi}}} + \underbrace{\begin{bmatrix} m_{R1}l_1d_1^\perp & \mathbf{0} \\ m_{R2}l_1d_1^\perp & m_{R2}l_2d_2^\perp\end{bmatrix}}_{A_{\ddot{\varphi}}}\ddot{\boldsymbol{\varphi}}$$

$$\mathbf{a}_{\mathbf{f}_L} = \underbrace{\begin{bmatrix} \mathbf{d}_1 & -\mathbf{d}_2 \\ \mathbf{0} & \mathbf{d}_2\end{bmatrix}}_{\mathbf{D}}\mathbf{f}_L,$$

where $\mathbf{a}_{\dot{\varphi}} \in \mathbb{R}^4$, $\mathbf{A}_{\ddot{\varphi}} \in \mathbb{R}^{4\times 2}$, $\mathbf{D} \in \mathbb{R}^{4\times 2}$ and $\mathbf{f}_L = [f_{L1} \;\; f_{L2}]^\top \in \mathbb{R}^2$. Therefore (7.2) can be rewritten as:

$$\underbrace{\begin{bmatrix}\mathbf{A}_{\ddot{\varphi}} & \mathbf{D}\end{bmatrix}}_{\mathbf{W}}\begin{bmatrix}\ddot{\boldsymbol{\varphi}} \\ \mathbf{f}_L\end{bmatrix} = -\mathbf{a}_{\mathbf{f}_R} - \mathbf{a}_{\mathbf{g}} - \mathbf{a}_{\dot{\varphi}}. \tag{7.4}$$

The matrix $\mathbf{W} \in \mathbb{R}^{4\times 4}$, that can be explicitly written as

$$\mathbf{W} = \begin{bmatrix} -l_1m_{R1}\sin\varphi_1 & 0 & \cos\varphi_1 & -\cos\varphi_2 \\ l_1m_{R1}\cos\varphi_1 & 0 & \sin\varphi_1 & -\sin\varphi_2 \\ -l_1m_{R2}\sin\varphi_1 & -l_2m_{R2}\sin\varphi_2 & 0 & \cos\varphi_2 \\ l_1m_{R2}\cos\varphi_1 & l_2m_{R2}\cos\varphi_2 & 0 & \sin\varphi_2\end{bmatrix},$$

is full rank, in fact its determinant is $\det(\mathbf{W}) = -l_1l_2m_{R2}[m_{R1} + m_{R2}(1 - \cos^2(\varphi_1 - \varphi_2))]$, which is always nonzero.

The dynamics of the system is then described by the following equations:

$$\ddot{\boldsymbol{\varphi}} = \begin{bmatrix}\mathbf{I}_2 & \mathbf{0}\end{bmatrix} W^{-1}(-\mathbf{a}_{\mathbf{f}_R} - \mathbf{a}_{\mathbf{g}} - \mathbf{a}_{\dot{\varphi}}). \tag{7.5a}$$

$$\ddot{\theta} = J^{-1}\tau_R \tag{7.5b}$$

For the design of a state observer in Sect. 7.4 it is useful to rewrite (7.5) in a Lagrangian format:

$$M(\varphi)\ddot{\varphi} = -c(\varphi, \dot{\varphi}) + \bar{\mathbf{Q}}_{\varphi}(\varphi, \theta)\mathbf{f}_R \tag{7.6a}$$

$$J\ddot{\theta} = \tau_R, \tag{7.6b}$$

where $\mathbf{f}_R = [f_{R1} \;\; f_{R2}]^\top$ and

$$M(\varphi) = \begin{bmatrix} (m_{R1} + m_{R2})l_1^2 & m_{R2}l_1l_2\cos(\varphi_1 - \varphi_2) \\ m_{R2}l_1l_2\cos(\varphi_1 - \varphi_2) & m_{R2}l_2^2 \end{bmatrix}$$

$$c(\varphi, \dot{\varphi}) = \begin{bmatrix} m_{R2}l_1l_2\sin(\varphi_1 - \varphi_2)\,\dot{\varphi}_2^2 + (m_{R1} + m_{R2})gl_1\cos\varphi_1 \\ -m_{R2}l_1l_2\sin(\varphi_1 - \varphi_2)\,\dot{\varphi}_1^2 + m_{R2}gl_2\cos\varphi_2 \end{bmatrix}$$

$$\bar{\mathbf{Q}}_{\varphi}(\varphi, \theta) = \begin{bmatrix} l_1\cos(\varphi_1 + \theta_1) & l_1\cos(\varphi_1 + \theta_2) \\ 0 & l_2\cos(\varphi_2 + \theta_2) \end{bmatrix}$$

7.2 Differential Flatness

For the single tethered aerial vehicle we showed that it is differentially flat with respect to $\mathbf{y}^a$ containing the position of the vehicle and the internal force along the link. In this section we shall show that analogously, the multi-robot extension here considered is differentially flat with respect to the output $\mathbf{y}_2^a = [\mathbf{y}_{21}^{a\,\top} \;\; \mathbf{y}_{22}^{a\,\top}]^\top = [\varphi^\top \;\; \mathbf{f}_L^\top]^\top$, containing the position of the vehicles (parametrized by the elevation angles) and the internal force along the links.

We recall that to prove the differential flatness of the system, state, $\mathbf{x} = [x_1 \;\; x_2 \;\; x_3 \;\; x_4 \;\; x_5 \;\; x_6 \;\; x_7 \;\; x_8]^\top = [\varphi_1 \;\; \dot{\varphi}_1 \;\; \varphi_2 \;\; \dot{\varphi}_2 \;\; \theta_1 \;\; \dot{\theta}_1 \;\; \theta_2 \;\; \dot{\theta}_2]^\top \in \mathbb{R}^8$ and input, $\mathbf{u} = [f_{R1} \;\; f_{R2} \;\; \tau_{R1} \;\; \tau_{R2}]^\top = [\mathbf{f}_R{}^\top \;\; \tau_R{}^\top]^\top = [\mathbf{u}_1^\top \;\; \mathbf{u}_2^\top]^\top \in \mathbb{R}^4$, have to be expressed as an algebraic function of the output and its derivatives. We have that φ is already part of the output, thus $\varphi = \mathbf{y}_{21}^a$ and $\dot{\varphi} = \dot{\mathbf{y}}_{21}^a$. To find the rest we firstly compute the nominal thrust vectors from the output and its derivatives using (7.2):

$$\begin{bmatrix} f_{R1}\mathbf{z}_{B_1} \\ f_{R2}\mathbf{z}_{B_2} \end{bmatrix} = \mathbf{a}(\mathbf{y}_{21}^a, \dot{\mathbf{y}}_{21}^a, \ddot{\mathbf{y}}_{21}^a) + \mathbf{a}_{\mathbf{f}_L}(\mathbf{y}_2^a) + \mathbf{a}_{\mathbf{g}}. \tag{7.7}$$

Similarly to Sect. 4.4.1, from the thrust vectors we can easily compute the inputs and the missing part of the state as function of the output and its derivatives up to the fourth derivative.

Proposition *The model (7.5), is differentially flat with respect to the flat output* $\mathbf{y}_2^a = [\varphi^\top \;\; \mathbf{f}_L^\top]^\top$. *In other words, the state and the inputs can be written as algebraic function of* $\mathbf{y}_2^a$ *and a finite number of its derivatives.* □

7.3 Dynamic Feedback Linearization

As usual, to compute the feedback linearizing control law, we need to differentiate the outputs until the input $\mathbf{u}$ appears. Inverting (7.4), and recalling that $\mathbf{y}_{21}^{a\,(2)} = \ddot{\boldsymbol{\varphi}}$ and $\mathbf{y}_{22}^{a} = \mathbf{f}_L$, we directly obtain

$$\begin{bmatrix} \mathbf{y}_{21}^{a\,(2)} \\ \mathbf{y}_{22}^{a} \end{bmatrix} = \underbrace{W^{-1}(-\mathbf{a}_g - \mathbf{a}_{\dot{\varphi}})}_{\mathbf{b}(\mathbf{x})} + \underbrace{\left(-W^{-1}\left[\mathbf{Z}_R\ \mathbf{0}\right]\right)}_{\mathbf{E}(\mathbf{x})}\mathbf{u}, \tag{7.8}$$

where $\mathbf{b}(\mathbf{x})$ gathers all the terms that do not depend on $\mathbf{u}$ and $\mathbf{Z}_R \in \mathbb{R}^{4\times 2}$ is:

$$\mathbf{a}_{\mathbf{f}_R} = \begin{bmatrix} \mathbf{z}_{B_1} & \mathbf{0} & \mathbf{0}\ \mathbf{0} \\ \mathbf{0} & \mathbf{z}_{B_2} & \mathbf{0}\ \mathbf{0} \end{bmatrix}\mathbf{u} = \left[\mathbf{Z}_R\ \mathbf{0}\right]\mathbf{u}. \tag{7.9}$$

From (7.8) we can see that the input appears directly in $\mathbf{y}_{22}^{a}$ without need for differentiation while $\mathbf{y}_{21}^{a}$ has to be differentiated twice. Furthermore, we can immediately notice that the decoupling matrix $\mathbf{E}(\mathbf{x})$ is always singular which means that it is not possible to determine a static feedback that linearizes the system using $\mathbf{y}_{2}^{a}$.

As we saw in Sect. 4.6, the common technique is to delay the appearance of the input in $\mathbf{y}_{22}^{a}$ (i.e., increasing the relative degree of $\mathbf{y}_{22}^{a}$) introducing a dynamic compensator composed by one or more integrators in the input channel $\mathbf{u}_1$. To this aim, we redefine the input as $\bar{\mathbf{u}} = [\ddot{\mathbf{u}}_1^\top\ \mathbf{u}_2^\top]^\top = [\bar{\mathbf{u}}_1^\top\ \bar{\mathbf{u}}_2^\top]^\top$, considering the acceleration of the thrust intensity as new controllable input, $\ddot{\mathbf{u}}_1 = \ddot{\mathbf{f}}_R$. The system is now described by the extended state $\bar{\mathbf{x}} = \begin{bmatrix} \boldsymbol{\varphi}^\top & \dot{\boldsymbol{\varphi}}^\top & \boldsymbol{\theta}^\top & \dot{\boldsymbol{\theta}}^\top & \mathbf{u}_1^\top & \dot{\mathbf{u}}_1^\top \end{bmatrix}^\top \in \mathbb{R}^{12}$, that contains also the thrusts and their derivatives. Considering the extended system and the new input, $\mathbf{y}_{21}^{a}$ and $\mathbf{y}_{22}^{a}$ have to be differentiated four and two times, respectively, in order to see the new input $\bar{\mathbf{u}}$ appear:

$$\begin{aligned} \begin{bmatrix} \mathbf{y}_{21}^{a\,(4)} \\ \mathbf{y}_{22}^{a\,(2)} \end{bmatrix} &= (\ddot{W}^{-1})(-\mathbf{a}_{\mathbf{f}_R} - \mathbf{a}_g - \mathbf{a}_{\dot{\varphi}}) + 2(\dot{W}^{-1})(-\dot{\mathbf{a}}_{\mathbf{f}_R} - \dot{\mathbf{a}}_{\dot{\varphi}}) + \\ &+ (W^{-1})(-\ddot{\mathbf{a}}_{\mathbf{f}_R} - \ddot{\mathbf{a}}_{\dot{\varphi}}). \end{aligned} \tag{7.10}$$

In the previous equation the inputs appear only in the term $\ddot{\mathbf{a}}_{\mathbf{f}_R}$ that can be rewritten as:

$$\ddot{\mathbf{a}}_{\mathbf{f}_R} = \ddot{\mathbf{a}}'_{\mathbf{f}_R}(\bar{\mathbf{x}}) + \ddot{\mathbf{A}}''_{\mathbf{f}_R}(\bar{\mathbf{x}})\bar{\mathbf{u}}, \tag{7.11}$$

where

$$\ddot{\mathbf{A}}''_{\mathbf{f}_R}(\bar{\mathbf{x}}) = \begin{bmatrix} -\sin\theta_1 & 0 & -\frac{f_{R1}\cos\theta_1}{J_{R1}} & 0 \\ \cos\theta_1 & 0 & -\frac{f_{R1}\sin\theta_1}{J_{R1}} & 0 \\ 0 & -\sin\theta_2 & 0 & -\frac{f_{R2}\cos\theta_2}{J_{R2}} \\ 0 & \cos\theta_2 & 0 & -\frac{f_{R2}\sin\theta_2}{J_{R2}} \end{bmatrix}. \tag{7.12}$$

We can compactly rewrite (7.10) as:

$$\begin{bmatrix} \mathbf{y}_{21}^{a\,(4)} \\ \mathbf{y}_{22}^{a\,(2)} \end{bmatrix} = \mathbf{b}(\bar{\mathbf{x}}) + \underbrace{\left(-\mathbf{W}^{-1}\ddot{\mathbf{A}}''_{\mathbf{f}_R}(\bar{\mathbf{x}})\right)}_{\bar{\mathbf{E}}(\bar{\mathbf{x}})}\bar{\mathbf{u}}, \tag{7.13}$$

where $\mathbf{b}(\bar{\mathbf{x}})$, whose expression is omitted here for the sake of brevity, collects all the terms in (7.10) that do not depend on the input. After some algebra, it is possible to analytically compute the determinant of the new decoupling matrix $\bar{\mathbf{E}}(\bar{\mathbf{x}})$:

$$\det\left(\bar{\mathbf{E}}(\bar{\mathbf{x}})\right) = -\frac{f_{R1}f_{R2}}{J_{R1}J_{R2}l_1l_2m_{R2}\left(m_{R1}+m_{R2}\sin^2(\varphi_1-\varphi_2)\right)},$$

which is zero iff $f_{R1}=0$ or $f_{R2}=0$ (same singularity of the single tether case). Therefore $\bar{\mathbf{E}}(\bar{\mathbf{x}})$ is always invertible except for the cases in which one of the two thrusts vanishes. Furthermore the total relative degree $r = 8+4 = 12$ is equal to the dimension of the extended state $\bar{\mathbf{x}}$. This means that the system does not have an internal dynamics, i.e., it is fully linearizable through dynamic feedback. In fact, designing the control input as

$$\bar{\mathbf{u}} = \bar{\mathbf{E}}^{-1}(\bar{\mathbf{x}})\left[-\mathbf{b}(\bar{\mathbf{x}})+\mathbf{v}\right], \tag{7.14}$$

where $\mathbf{v} = [\mathbf{v}_1^\top \; \mathbf{v}_2^\top]^\top \in \mathbb{R}^4$ is a virtual input, we obtain

$$\mathbf{y}_{21}^{a\,(4)} = \mathbf{v}_1 \qquad \mathbf{y}_{22}^{a\,(2)} = \mathbf{v}_2,$$

i.e., through the state feedback transformation (7.14) we transform the original non linear system (7.5) in a fully-equivalent linear and decoupled dynamical system.

Proposition *Consider the system composed by two aerial vehicles connected in series to the ground by two links with passive joints, whose dynamic model is described by (7.5). Consider as outputs the elevation and the internal force of the two links,* $\mathbf{y}_2^a = [\boldsymbol{\varphi}^\top \; \mathbf{f}_L^\top]^\top$. *Then the system is fully linearizable via dynamic feedback for every state configuration, iff both thrusts* f_{R1} *and* f_{R2} *are nonzero.* □

As a consequence of the previous Proposition 8, as done in Sect. 4.6 we can design a standard linear controller to obtain the tracking of a desired trajectory. The overall controller design is depicted in Fig. 7.2.

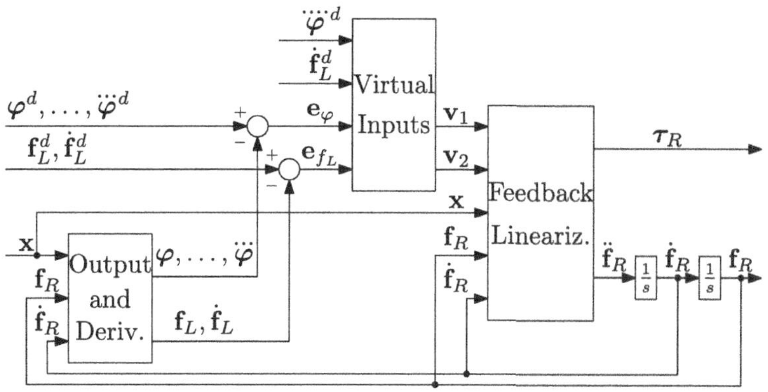

Fig. 7.2 Graphic representation of the controller. © 2020 IEEE. Reprinted, with permission, from [2]

Corollary *Let be given any desired trajectory $\mathbf{y}_{21}^{a\;d}(t)$ of class C^3 for the two links elevation $\mathbf{y}_{21}^a$, and any desired trajectory $\mathbf{y}_{22}^{a\;d}(t)$ of class C^1 for the two links internal force $\mathbf{y}_{22}^a$. Consider as input the second derivative of the two thrusts and the torques provided by the aerial vehicles, $\bar{\mathbf{u}} = [\ddot{\mathbf{f}}_R^\top \;\; \tau_R{}^\top]^\top$. Consider the control law described by (7.14) and set the virtual inputs as:*

$$\begin{aligned} \mathbf{v}_1 &= \mathbf{y}_{21}^{a\;d(4)} + \mathbf{K}_{11}\boldsymbol{e}_1 + \mathbf{K}_{12}\boldsymbol{e}_1^{(1)} + \mathbf{K}_{13}\boldsymbol{e}_1^{(2)} + \mathbf{K}_{14}\boldsymbol{e}_1^{(3)} \\ \mathbf{v}_2 &= \mathbf{y}_{22}^{a\;d(2)} + \mathbf{K}_{21}\boldsymbol{e}_2 + \mathbf{K}_{22}\boldsymbol{e}_2^{(1)}, \end{aligned} \tag{7.15}$$

where $\mathbf{K}_{ij} \in \mathbb{R}^{2\times 2}_{>0}$, with $i = 1\dots 4$ and $j = 1, 2$, are diagonal matrices.

That control law exponentially steers $\mathbf{y}_2^a$ along any desired trajectory $\mathbf{y}_2^{ad} = [\varphi_1^d \;\; \varphi_2^d \;\; f_{L1}^{\;d} \;\; f_{L2}^{\;d}]^\top$. The behavior of the convergence can be arbitrarily assigned by suitably choosing the gain matrixes. □

Let us define the errors as $\boldsymbol{e}_1 = \mathbf{y}_{21}^{a\;d} - \mathbf{y}_{21}^a$ and $\boldsymbol{e}_2 = \mathbf{y}_{22}^{a\;d} - \mathbf{y}_{22}^a$. The controller yields to the following error dynamics:

$$\begin{aligned} &\boldsymbol{e}_1^{(4)} + \mathbf{K}_{11}\boldsymbol{e}_1 + \mathbf{K}_{12}\boldsymbol{e}_1^{(1)} + \mathbf{K}_{13}\boldsymbol{e}_1^{(2)} + \mathbf{K}_{14}\boldsymbol{e}_1^{(3)} = 0 \\ &\boldsymbol{e}_2^{(2)} + \mathbf{K}_{21}\boldsymbol{e}_2 + \mathbf{K}_{22}\boldsymbol{e}_2^{(1)} = 0. \end{aligned}$$

Therefore, from basic linear system theory, one can arbitrarily assign the poles of the dynamics of the error in order to guaranties an arbitrarily fast exponential tracking of $(\mathbf{y}_{21}^{a\;d}(t), \mathbf{y}_{22}^{a\;d}(t))$ for $(\mathbf{y}_{21}^a(t), \mathbf{y}_{22}^a(t))$ by suitably choosing the gains: $\bar{\mathbf{K}}_1 = [\mathbf{K}_{11} \;\; \mathbf{K}_{12} \;\; \mathbf{K}_{13} \;\; \mathbf{K}_{14}] \in \mathbb{R}^{2\times 8}_{>0}$ and $\bar{\mathbf{K}}_2 = [\mathbf{K}_{21} \;\; \mathbf{K}_{22}] \in \mathbb{R}^{2\times 4}_{>0}$. Since $(\mathbf{y}_{21}^{a\;d(4)}(t), \mathbf{y}_{22}^{a\;d(2)}(t))$ have to be well defined, the elevation and internal force trajectories have to be of class C^3 and C^1 respectively.

Due to Proposition 8, if the links are bars, it is feasible to pass from compression to tension and viceversa. Instead, in the case of a cable, it is possible to maintain a sufficient value of tension under a maximum breaking value and above the minimum tautness value.

We remark that, since the total relative degree is equal to the dimension of the extended state, there is no internal dynamics. This implies that the dynamics of the pitch of each vehicles is stable during the tracking of the desired output.

Remark (*Case of zero thrust*) If a particular desired trajectory of the outputs requires zero thrust on one of the two vehicles the controller cannot be applied, indeed in this case it has a singularity. Thus, this fact has to be considered in the planning phase in order to design desired trajectories that ensure strictly positive, or negative, thrusts. Although this is a planning problem that does not concern this work, we believe that the problem of zero thrust does not imply a strong limitation on the set of the feasible trajectories. Indeed, as it is shown in Sect. 7.5, we can still generate non-trivial trajectories, e.g., inversion of the internal link force from tension to compression, ensuring non zero thrusts. An extended study on the planning of feasible trajectories is left as future work. □

Looking at the control law described by the Eqs. (7.14) and (7.15), and depicted in the block diagram of Fig. 7.2, one can notice that its implementation requires the knowledge of the extended state $\bar{\mathbf{x}}$, the output $\mathbf{y}$ and its derivatives (up to the third-order for $\mathbf{y}_{21}^a$ and first-order for $\mathbf{y}_{22}^a$). Nevertheless, $\mathbf{y}$ and all its needed derivatives can be calculated as function of $\mathbf{x}$ and $\bar{\mathbf{u}}$ as done, e.g., in (7.8) and (7.13) for some of the derivatives. Note also that $\mathbf{u}_1$ and $\dot{\mathbf{u}}_1$ are directly known because they are internal state of the controller.

7.4 State Estimation

Instead of considering a direct measure of the state to close the control loop, we aim to find the minimal set of sensors based on which we can obtain an estimation of the state. Inspired by the results obtained in Chap. 4 we consider the possible sensory setup of Table 7.1. In this section, for the case 4 of Table 7.1, we present a method to transform the original measurements into direct measurements of the configuration $\mathbf{q}$ and we show that this implies the observability of the full state, i.e., $\mathbf{q}$ and $\dot{\mathbf{q}}$. For this case we propose a nonlinear estimator, based on the HGO able to retrieve the state from any dynamic condition. In the end we analyze the applicability of the method to the other configurations of Table 7.1.

7.4.1 Output Transformations

Assume to have an onboard accelerometer for each robot, placed at O_{B_i} and attached to $\mathcal{F}_{Bi}$. According to the model in Sect. 3.4, it measures the specific acceleration:

Table 7.1 Possible sensors configurations. The $1st$ sensor type corresponds to an accelerometer mounted on each robot

Case	$2nd$ Sensor type	Mounting place	Measures ρ_1, ρ_2	Applicability
1	Absolute Inclinom.	$\mathcal{F}_{B1}$	θ_1	Yes
	Absolute Inclinom.	$\mathcal{F}_{B2}$	θ_2	
2	Absolute Inclinom.	$\mathcal{F}_{B1}$	θ_1	Yes
	Relative Inclinom.	$\mathcal{F}_{B2}$	$\theta_1 - \theta_2$	
3	Encoder	$\mathcal{F}_W - link_1$	φ_1	No
	Encoder	$\mathcal{F}_{B1} - link_1$	$\varphi_1 + \theta_1$	
4	Encoder	$\mathcal{F}_W - link_1$	φ_1	Yes
	Encoder	$\mathcal{F}_{B1} - link_2$	$\varphi_2 + \theta_1$	
5	Encoder	$\mathcal{F}_W - link_1$	φ_1	No
	Encoder	$\mathcal{F}_{B2} - link_2$	$\varphi_2 + \theta_2$	
6	Encoder	$\mathcal{F}_{B1} - link_1$	$\varphi_1 + \theta_1$	No
	Encoder	$\mathcal{F}_{B1} - link_2$	$\varphi_2 + \theta_1$	
7	Encoder	$\mathcal{F}_{B1} - link_1$	$\varphi_1 + \theta_1$	No
	Encoder	$\mathcal{F}_{B2} - link_2$	$\varphi_2 + \theta_2$	

$$\mathbf{a}_i = R_W^{B_i}(\ddot{\mathbf{p}}_{B_i} + g\mathbf{z}_W) = \left[a_{i_x}\ 0\ a_{i_z}\right]^\top, \tag{7.16}$$

where $R_W^{B_i} \in \mathbb{R}^3$ is the rotation matrix from $\mathcal{F}_W$ to $\mathcal{F}_{Bi}$, and $\ddot{\mathbf{p}}_{B_i}$ is the acceleration of the CoM of the ith vehicle w.r.t. $\mathcal{F}_W$.

Then we assume to be in the case #4 of Table 7.1, i.e., the system is equipped with two encoders, one is rigidly attached to the ground and connected to the first link and measures its absolute elevation relative to $\mathcal{F}_W$, while the second is fixed to $\mathcal{F}_{B1}$ and connected to the second link, and measures its relative elevation with respect to $\mathcal{F}_{B1}$, i.e.:

$$\rho_1 = \varphi_1, \qquad \rho_2 = \varphi_2 + \theta_1. \tag{7.17}$$

Now, replacing $\ddot{\mathbf{p}}_{B_2}$ from (7.2) into (7.16) for $i = 2$, we obtain

$$-m_{R2}\mathbf{a}_2 = R_W^{B_i}(f_{L2}\mathbf{d}_2 + f_{R2}\mathbf{z}_{B_2}) = f_{L2}R_W^{B_i}\mathbf{d}_2 + [0\ 0\ f_{R2}]^\top, \tag{7.18}$$

which allows to define the measurement transformation

$$\begin{bmatrix} w_1(k) \\ w_2(k) \end{bmatrix} = \begin{bmatrix} \operatorname{sgn}\left(k - \frac{1}{2}\right)\sqrt{\bar{a}_{2_x}^2 + \bar{a}_{2_z}^2} \\ \operatorname{atan2}\left(\dfrac{\bar{a}_{2_z}}{w_1(k)}, \dfrac{\bar{a}_{2_x}}{w_1(k)}\right) \end{bmatrix} = \begin{bmatrix} 0 \\ \varphi_2 + \theta_2 + \frac{\pi}{2} \end{bmatrix} \pm \begin{bmatrix} f_{L2} \\ \frac{\pi}{2} \end{bmatrix}, \tag{7.19}$$

where $\bar{a}_{2_x} = m_{R2}a_{2_x}$, $\bar{a}_{2_z} = m_{R2}a_{2_z} + f_{R2}$ and $k \in \{0, 1\}$. Note that [*(i)*] there are two solutions for $k = 0$ and $k = 1$ because $\operatorname{sgn}(f_{L2})$ is not retrievable from the measurements; [*(ii)*]the transformation is allowed iff $f_{L2} \neq 0$.

At every time instant t there is only one *correct* pair of measurements, equal to $(f_{L2}, \varphi_2 + \theta_2)$, while the other is *wrong* and equal to $(-f_{L2}, \varphi_2 + \theta_2 + \pi)$. We define k^* the unique $k \in \{0, 1\}$ such that $(w_1(k^*), w_2(k^*)) = (f_{L2}, \varphi_2 + \theta_2)$. Then, replacing $\ddot{\mathbf{p}}_{B_1}$ from (7.2) into (7.16) for $i = 1$, and after some simple algebra, we can define two additional new measurement transformations:

$$\begin{bmatrix} w_3(k^*, j) \\ w_4(k^*, j) \end{bmatrix} = \begin{bmatrix} \operatorname{sgn}\left(j - \frac{1}{2}\right)\sqrt{\bar{a}_{1_x}^2 + \bar{a}_{1_z}^2} \\ \operatorname{atan2}\left(\dfrac{\bar{a}_{1_z}}{w_3(k^*, j)}, \dfrac{\bar{a}_{1_x}}{w_3(k^*, j)}\right) \end{bmatrix} = \begin{bmatrix} 0 \\ \varphi_1 + \theta_1 + \frac{\pi}{2} \end{bmatrix} \pm \begin{bmatrix} f_{L1} \\ -\frac{\pi}{2} \end{bmatrix}, \tag{7.20}$$

where $\bar{a}_{1_x} = m_{R1}a_{1_x} - w_1(k^*)\cos\rho_2$, $\bar{a}_{1_z} = m_{R2}a_{1_z} - w_1(k^*)\sin\rho_2 + f_{R2}$ and $j \in \{0, 1\}$. As in (7.19), the transformation is not possible when $f_{L1} = 0$. A practical solution for the instantaneous zero internal force case is provided in Sect. 7.4.5.

Since the sign of f_{L1} is not retrievable from the measurements, we obtain two solutions parametrized by j, i.e., $(w_3(k^*, j), w_4(k^*, j))$. At every time instant t there is only one *correct* pair of measurements equal to $(f_{L1}, \varphi_1 + \theta_1)$, while the other is *wrong* and equal to $(-f_{L1}, \varphi_1 + \theta_1 + \pi)$. Actually, recalling that also $k \in \{0, 1\}$, we obtain four groups of different measurements, i.e., $(w_1(k), w_2(k), w_3(k, j), w_4(k, j))$ with $k, j \in \{0, 1\}$. We know that at each time t there is only one couple $k^*, j^* \in \{0, 1\}$ such that the corresponding measurements are *correct*, i.e., $(w_1(k^*), w_2(k^*), w_3(k^*, j^*), w_4(k^*, j^*)) = (f_{L2}, \varphi_2 + \theta_2, f_{L1}, \varphi_1 + \theta_1)$, while all the others are *wrong*.

Finally, exploiting the readings of the encoders, we can define the last measurement transformation

$$\begin{aligned} \eta_1 &= \rho_1 \\ \eta_2(k, j) &= \rho_1 + \rho_2 - w_4(k, j) \\ \eta_3(k, j) &= w_4(k, j) - \rho_1 \\ \eta_4(k, j) &= w_2(k) + w_4(k, j) - \rho_1 - \rho_2. \end{aligned} \tag{7.21}$$

The transformation method is represented in Fig. 7.3a. From (7.21) one can notice that for the pair (k^*, j^*) defined before, we obtain a direct measure of the generalized coordinates, i.e. $\boldsymbol{\eta}(k^*, j^*) = [\eta_1 \ \eta_2(k^*, j^*) \ \eta_3(k^*, j^*) \ \eta_4(k^*, j^*)]^\top = [\boldsymbol{\eta}_1^\top(k^*, j^*) \ \boldsymbol{\eta}_2^\top(k^*, j^*)]^\top = [\varphi_1 \ \varphi_2 \ \theta_1 \ \theta_2]^\top$. While, for the pairs $(k, j) \neq (k^*, j^*)$, $\boldsymbol{\eta}(k, j)$ is a wrong measurement of the configuration. From a single set of measures it is not possible to discriminate which is the correct pair (k^*, j^*) corresponding to the correct $\boldsymbol{\eta}$, nevertheless, in Sect. 7.4.4 we show a discriminating method exploiting the dynamics, similar to the one proposed in Sect. 4.8.

For the purpose of proving the observability of the system and for designing the observer we consider $\boldsymbol{\eta} = \boldsymbol{\eta}(k^*, j^*)$.

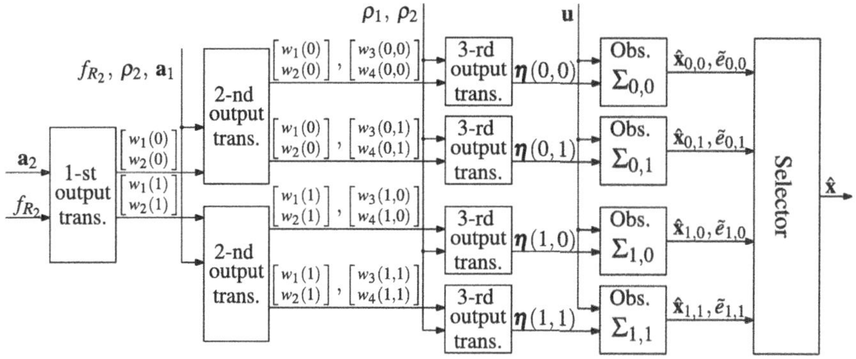

(a) Global Observer.

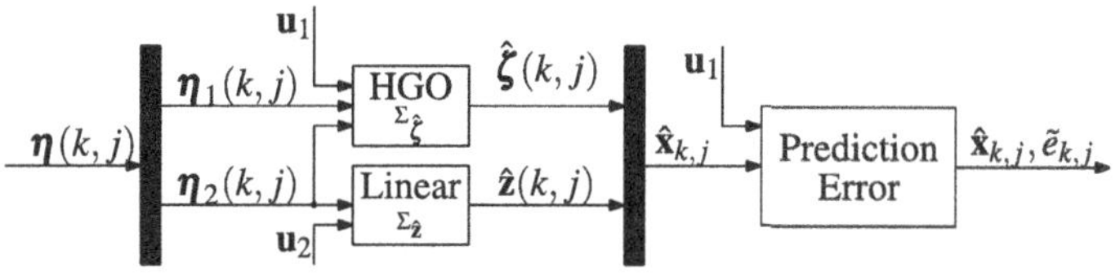

(b) Observer $\Sigma_{k,j}$.

Fig. 7.3 Graphic representation of the observer. © 2020 IEEE. Reprinted, with permission, from [1]

7.4.2 *Observability*

In order to study the observability of the system and to design an observer of the state, we first rewrite the system in a state space form. We can rewrite (7.6) and the measurements function (7.21) as:

$$\dot{\mathbf{x}} = \mathbf{A}\mathbf{x} + \mathbf{B}\begin{bmatrix}\boldsymbol{\Sigma}(\mathbf{x}, \mathbf{u}_1) \\ J^{-1}\mathbf{u}_2\end{bmatrix} \tag{7.22a}$$

$$\eta = \mathbf{C}\mathbf{x}, \tag{7.22b}$$

where $\mathbf{A} = \text{diag}(\mathbf{A}_1, \mathbf{A}_2, \mathbf{A}_3, \mathbf{A}_4)$, $\mathbf{B} = \text{diag}(\mathbf{B}_1, \mathbf{B}_2, \mathbf{B}_3, \mathbf{B}_4)$, $\mathbf{C} = \text{diag}(\mathbf{C}_1, \mathbf{C}_2, \mathbf{C}_3, \mathbf{C}_4)$ and

$$\mathbf{A}_i = \begin{bmatrix}0 & 1 \\ 0 & 0\end{bmatrix}, \quad \mathbf{B}_i = \begin{bmatrix}0 \\ 1\end{bmatrix}, \quad \mathbf{C}_i = \begin{bmatrix}1 & 0\end{bmatrix} \quad \forall i = 1, \dots, 4$$

$$\boldsymbol{\Sigma}(\mathbf{x}, \mathbf{u}_1) = -\boldsymbol{M}(\mathbf{x})^{-1}\boldsymbol{c}(\mathbf{x}) + \boldsymbol{M}(\mathbf{x})^{-1}\bar{\mathbf{Q}}_{\varphi}(\mathbf{x})\boldsymbol{u}_1. \tag{7.23}$$

Notice that $\boldsymbol{M}(\mathbf{x})$ is always invertible. Writing (7.22) as $\dot{\mathbf{x}} = \mathbf{f}(\mathbf{x}, \mathbf{u})$, and $\eta = \mathbf{h}(\mathbf{x})$. the system results observable if the nonlinear observability matrix $O(\mathbf{x}, \mathbf{u}) =$

$\left[\frac{\partial \mathbf{h}(\mathbf{x})}{\partial \mathbf{x}}, \frac{\partial \dot{\mathbf{h}}(\mathbf{x})}{\partial \mathbf{x}}, \ldots, \frac{\partial \mathbf{h}^{(7)}(\mathbf{x})}{\partial \mathbf{x}}\right]^\top \in \mathbb{R}^{4\cdot 8\times 8}$ is full rank [3]. We can notice that

$$O(\mathbf{x}, \mathbf{u})_1 = \left[\frac{\partial \mathbf{h}(\mathbf{x})}{\partial \mathbf{x}}, \frac{\partial \dot{\mathbf{h}}(\mathbf{x})}{\partial \mathbf{x}}\right]^\top = \left[\mathbf{C}^\top \ (\mathbf{C}A)^\top\right]^\top.$$

Changing the order of the rows we obtain $O(\mathbf{x}, \mathbf{u})'_1 = \mathbf{I}_8$, that is full rank for every $\mathbf{x} \in \mathbb{R}^8$ and $\mathbf{u} \in \mathbb{R}^4$. This implies that also $O(\mathbf{x}, \mathbf{u})$ is always full rank, i.e.,

Proposition *Consider the system described by (7.6) with two on-board accelerometers, mounted on each vehicles, and two encoders. One is attached to the ground and connected to the first link, and one is mounted on the first vehicle and connected to the second link. Then, the system is observable except for the zero internal force cases, i.e., $f_{L1} = 0$ or $f_{L2} = 0$.* □

Although we proved Proposition 7.4.2 only for the fourth case of Table 7.1, actually, the result shows a more general sufficient observability condition. Indeed, independently from the available sensors, whenever there are some output transformations that translate the original measurements into direct measures of $\mathbf{q}$, then the system is observable, i.e.,

Proposition *Consider the system described by (7.22a) and a set of measurements $\mathbf{w} = \mathbf{h}(\mathbf{x}, \mathbf{u}) \in \mathbb{R}^p$, where $p \in \mathbb{R}_{\geq 1}$. Define $\mathcal{X}$ the state space and $\mathcal{U}$ the control inputs space. If there exists a subspace $\mathcal{D} \subseteq \mathcal{X} \times \mathcal{U}$ and a measurement transformation function $\mathbf{\Gamma} : \mathbb{R}^p \to \mathbb{R}^4$ valid in $\mathcal{D}$, such that $[\varphi_1 \ \varphi_2 \ \theta_1 \ \theta_2]^\top = \mathbf{\Gamma}(\mathbf{w})$, then the system is observable for every $\mathbf{x}$ and $\mathbf{u}$ in $\mathcal{D}$, and can be written in the form of (7.22).* □

7.4.3 High Gain Observer

For the sets of measurements that fulfill the condition of Proposition 7.4.2, and in particular for the case 4 of Table 7.1 we show in this section the design of an observer based on HGO (see Sect. 2.4).

Considering the system (7.22) we define $\boldsymbol{\zeta} = [\boldsymbol{\zeta}_1^\top \ \boldsymbol{\zeta}_2^\top]^\top = [\zeta_1 \ \zeta_2 \ \zeta_3 \ \zeta_4]^\top = [x_1 \ x_2 \ x_3 \ x_4]^\top$ and $\mathbf{z} = [\mathbf{z}_1^\top \ \mathbf{z}_2^\top]^\top = [z_1 \ z_2 \ z_3 \ z_4]^\top = [x_5 \ x_6 \ x_7 \ x_8]^\top$. The system (7.22) can be then written as

$$\begin{cases} \dot{\boldsymbol{\zeta}} = \mathbf{A}_\zeta \boldsymbol{\zeta} + \mathbf{B}_\zeta \mathbf{\Sigma}(\boldsymbol{\zeta}, \eta_2, \mathbf{u}_1) \\ \eta_1 = \mathbf{C}_\zeta \boldsymbol{\zeta} \end{cases} \qquad \begin{cases} \dot{\mathbf{z}} = \mathbf{A}_\mathbf{z} \mathbf{z} + \mathbf{B}_\mathbf{z} \mathbf{u}_2 \\ \eta_2 = \mathbf{C}_\mathbf{z} \mathbf{z}, \end{cases} \tag{7.24}$$

where $\mathbf{A}_\zeta = \mathrm{diag}(\mathbf{A}_1, \mathbf{A}_2)$, $\mathbf{B}_\zeta = \mathrm{diag}(\mathbf{B}_1, \mathbf{B}_2)$, $\mathbf{C}_\zeta = \mathrm{diag}(\mathbf{C}_1, \mathbf{C}_2)$, $\mathbf{A}_\mathbf{z} = \mathrm{diag}(\mathbf{A}_3, \mathbf{A}_4)$, $\mathbf{B}_\mathbf{z} = \mathrm{diag}(\mathbf{B}_3, \mathbf{B}_4)J^{-1}$, $\mathbf{C}_\mathbf{z} = \mathrm{diag}(\mathbf{C}_3, \mathbf{C}_4)$. Having replaced θ_1 and θ_2 with their measures $\boldsymbol{\eta}_2$ in the dynamics of $\boldsymbol{\zeta}$, the two systems become completely independent, moreover, the second one is linear, therefore we can design for it a classical Luenberger observer

$$\dot{\hat{\mathbf{z}}} = \mathbf{A_z}\hat{\mathbf{z}} + \mathbf{B_z}\mathbf{u}_2 + \mathbf{H_z}(\eta_2 - \mathbf{C_z}\hat{\mathbf{z}}), \tag{7.25}$$

where $\mathbf{H_z} = \text{diag}(\mathbf{H}_{\mathbf{z}_1}, \mathbf{H}_{\mathbf{z}_1})$ and $\mathbf{H}_{\mathbf{z}_i} = [\beta_1^i \ \beta_2^i]^\top$, whose elements, $\beta_j^i \in \mathbb{R}_{>0}$ can be set to place the poles of the error dynamics, $\boldsymbol{e}_{\mathbf{z}_i} = \mathbf{z}_i - \hat{\mathbf{z}}_i$. Instead, for the first system, thanks to its particular triangular form, it is possible to use the following HGO

$$\dot{\hat{\zeta}} = \mathbf{A}_\zeta\hat{\zeta} + \mathbf{B}_\zeta\mathbf{\Sigma}(\hat{\zeta}, \eta_2, \mathbf{u}_1) + \mathbf{H}_\zeta(\eta_1 - \mathbf{C}_\zeta\hat{\zeta}), \tag{7.26}$$

where $\mathbf{H}_\zeta = \text{diag}(\mathbf{H}_{\zeta_1}, \mathbf{H}_{\zeta_1})$ and $\mathbf{H}_{\zeta_i} = [\frac{\alpha_1^i}{\epsilon} \ \frac{\alpha_2^i}{\epsilon^2}]^\top$, with $\epsilon \in \mathbb{R}_{>0}$, and the gains $\alpha_j^i \in \mathbb{R}_{>0}$ are set such that the roots of $s^2 + \alpha_1^i s + \alpha_2^i$ have negative real part. The gains (α_1^i, α_2^i) influence the convergence rate of the estimation of the ith elevation angle and its derivative, i.e., φ_i and $\dot{\varphi}_i$. A schematic representation of the observer is given in Fig. 7.3b.

7.4.4 Disambiguation of η

The output transformations described in Sect. 7.4.1 generates four different set of measurements, $\eta(k, j)$ with $k, j \in \{0, 1\}$, of which only one is correct.

As represented in Fig. 7.3a, for each $k, j \in \{0, 1\}$, we implement an observer of the state, Σ_{kj}, using (7.25) and (7.26), based on the measurements $\eta(k, j)$. Therefore we obtain four estimates of the state, one for each measurement pair, $\hat{\mathbf{x}}_{0,0}$, $\hat{\mathbf{x}}_{0,1}$, $\hat{\mathbf{x}}_{1,0}$, $\hat{\mathbf{x}}_{1,1}$, and the correct one has to be recognized.

Define $\hat{\mathbf{w}} = [\hat{\mathbf{a}}_1^\top \ \hat{\mathbf{a}}_2^\top \ \hat{\rho}_1 \ \hat{\rho}_2]^\top$ as the vector that contains the measurements computed with the estimated state, i.e.,

$$\hat{\rho}_1 = \hat{x}_1, \quad \hat{\rho}_2 = \hat{x}_3 + \hat{x}_5, \quad \hat{\mathbf{a}}_i = \hat{R}_W^{B_i}(\hat{\ddot{\mathbf{p}}}_{B_i} - g z_W),$$

where $\hat{R}_W^{B_i} = R_W^{B_i}(\hat{\mathbf{x}})$, and $\hat{\ddot{\mathbf{p}}}_{B_i}$ is calculated considering the system model (i.e., no numerical differentiation is needed)

$$\hat{\ddot{\mathbf{p}}}_{B_1} = -l_1\mathbf{d}_1(\hat{x}_1)\hat{x}_2^2 + l_1\mathbf{d}_1^\perp(\hat{x}_1)\hat{\dot{x}}_2$$
$$\hat{\ddot{\mathbf{p}}}_{B_2} = \hat{\ddot{\mathbf{p}}}_{B_1} - l_2\mathbf{d}_2(\hat{x}_3)\hat{x}_4^2 + l_2\mathbf{d}_2^\perp(\hat{x}_3)\hat{\dot{x}}_4.$$

In the previous equations $\hat{\dot{x}}_2$ and $\hat{\dot{x}}_4$ are the estimation of the angular acceleration of the elevations calculated replacing the estimated state into (7.23), i.e., $[\hat{\dot{x}}_2 \ \hat{\dot{x}}_4]^\top = \mathbf{\Sigma}(\hat{\mathbf{x}}, \mathbf{u}_1)$ (no numerical differentiation needed in this case either).

In order to choose the correct estimation among the four, we propose a method based on the minimal prediction error, similar to the one used in Sect. 4.8. For each observer we compute a prediction error $\tilde{e}_{k,j}$ smoothed with an exponential discount factor:

$$\dot{\tilde{e}}_{k,j} = \lambda(\|\mathbf{w} - \hat{\mathbf{w}}_{k,j}\| - \tilde{e}_{k,j}),$$

where $\lambda \in \mathbb{R}_{>0}$ sets the discount rate and $\mathbf{w} = [\mathbf{a}_1^\top \ \mathbf{a}_2^\top \ \rho_1 \ \rho_2]^\top$. Then, the estimation of the observer with minimum prediction error is chosen, i.e., $\hat{\mathbf{x}} = \hat{\mathbf{x}}_{k^*,j^*}$ s.t. $\{k^*, j^*\} = \arg\min_{\mathrm{k,j} \in \{0,1\}}(\tilde{e}_{k,j})$. Figure 7.3 shows the estimator structure.

7.4.5 *Discussion on the Proposed Method*

7.4.5.1 Zero Internal Force Case

As we previously noticed, if one of the link internal forces is zero then w_2 or w_4 cannot be determined. We noticed an analogous singularity in Sects. 4.7 and 4.8 for the single tethered system as well. Nevertheless, we showed that if the desired internal force is passing through zero for a sufficiently short (ideally zero) time interval, one can still use the proposed observes in practice by updating the filter without the correction term in that time instants. For the multi-tethered system this implies to impose

$$\begin{aligned} \dot{\hat{\mathbf{z}}} &= \mathbf{A}_z\hat{\mathbf{z}} + \mathbf{B}_z\mathbf{u}_2 \\ \dot{\hat{\boldsymbol{\xi}}} &= \mathbf{A}_\xi\hat{\boldsymbol{\xi}} + \mathbf{B}_\xi\boldsymbol{\Sigma}(\hat{\boldsymbol{\xi}}, \hat{\mathbf{z}}, \mathbf{u}_1) \end{aligned} \qquad \text{if } w_1 = 0 \text{ or } w_3 = 0.$$

During this instant the observation is done in 'open loop' only using the model dynamics, thus the error dynamics becomes non strictly stable for a short moment. However, the dynamics returns asymptotically stable as soon as the internal force becomes non-zero again, as it is shown in Sect. 7.5 by simulations.

7.4.5.2 Applicability

The transformation method showed for the case 4 in Table 7.1 can be applied also to other sets of sensors. Last column of Table 7.1 specifies for which cases the method is able to transform the original measurements into direct measures of the system configuration. For cases 1 and 2, the measurements transformations are very similar to those derived for case 4. For the remaining cases it is not possible to apply the proposed method. In particular for the cases 3, 5, 7 we cannot compute the transformation (7.20). While, for the case 6, the problem lies in the last transformation (7.21).

7.4.5.3 Loop Stability

For the control law described in Sect. 7.3, the knowledge of the state is sufficient in order to close the loop. Thus we can use as feedback the state estimation provided by the proposed observer. Then a similar reasoning to the one in Sects. 4.7 and 4.8

can be done to prove that there exist a ϵ^* such that, for every $0 < \epsilon \leq \epsilon^*$ in (7.26), the closed loop system with the observer is exponentially stable, except for the zero thrust and zero internal force cases.

7.5 Numerical Validation

We tested the closed loop system (observer + controller) in simulation using two aerial robots with $m_{Ri} = 1\,\text{kg}$ and $J_{Ri} = 0.15\,\text{kg m}^2$, and two links with $l_i = 2\,\text{m}$ $(i = 1, 2)$. In order to obtain a reasonable fast tracking of the desired trajectories we set the gains such as the error dynamics relative to φ_1, φ_2 and f_{L1}, f_{L2} has poles in $(-3, -6, -9, -12)$ and $(-5, -10)$, respectively.

Regarding the convergence of the state estimation, we set $\epsilon = 0.1$ and the gains (α_1^i, α_2^i) such as the roots of $s^2 + \alpha_1^i s + \alpha_2^i$ are $(-2, -3)$. We set $\mathbf{H}_{\mathbf{z}_i}$ such that the error dynamics of the estimation of θ_1 and θ_2 has poles in $(-15, -25)$. Finally, the discount rate of the prediction error dynamics is set to $\lambda = 20$. These gains values, replicated identically for each of the four observers, guarantee the stability of the closed-loop system.

To show the ability of the proposed observer to exponentially converge to the real state, we initialize it with an error of 5° relatively to the elevation and pitch angles. We propose two different simulations:

(i) the first, whose results are plotted in Fig. 7.4, shows the performances of the global closed loop system in the particular case of inversion of the internal force. In particular, the trajectory of the end-effector is a trajectory of class C^3 from the initial position $\mathbf{p}_{B_2}(0) = [2.5 \;\, 0 \;\, 2]^T$ to the final $\mathbf{p}_{B_2}(t_f) = [-0.7 \;\, 0 \;\, 0.7]^T$. While the desired internal force along the links is a trajectory of class C^1 from the initial tension of 10 N to the final compression of -10 N.

(ii) In the second simulation, reported in Fig. 7.5, we replicate a plausible real scenario where the system is controlled as a two-link robot. The desired trajectory of the end effector is planned in the Cartesian space as a sequence of three arcs of ellipse in order to enter, stop on each room of a plausible building, and then return to the initial position. By inverse kinematics the desired trajectories of the two elevations are derived. In the meanwhile a constant tension of 5 N is required on the two links for the hole duration of the task.

To better represent the behavior of the system, Figs. 7.4c and 7.5c show the stroboscopic evolution of the system where the flow of time is provided by the change of color. To graphically represent the internal force variation, the link is drawn as a dashed line with a thinner width when the tension is higher, and as a solid line with a wider width when the compression is higher.

From Figs. 7.4 and 7.5 one can notice that the estimation of the state converges to the real one in less than one second, in any dynamic condition. Moreover, for the first simulation, the prediction error does not increase even when the desired internal force passes through zero. Although during the transient of the estimation the controller

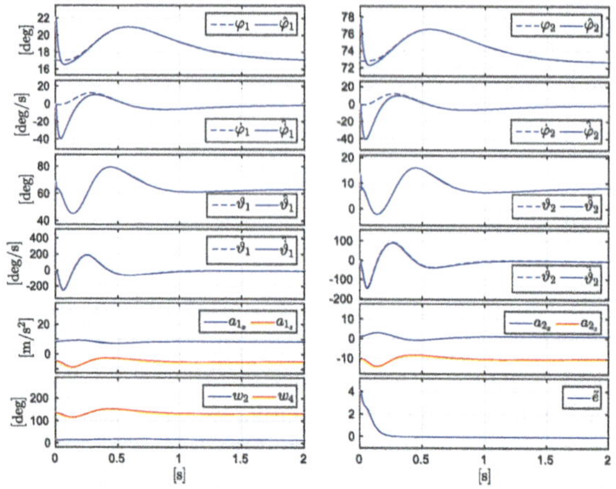

(a) Observer Results: the plot shows only the first 2 seconds of simulation. After this time the estimation follows the state with high fidelity thus the rest of the simulation is not displayed.

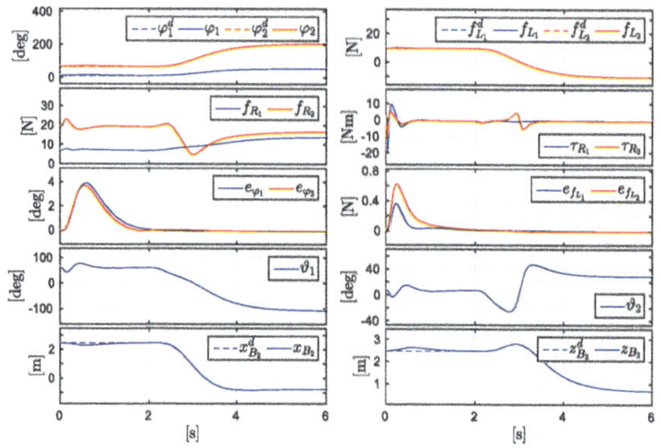

(b) Controller Results.

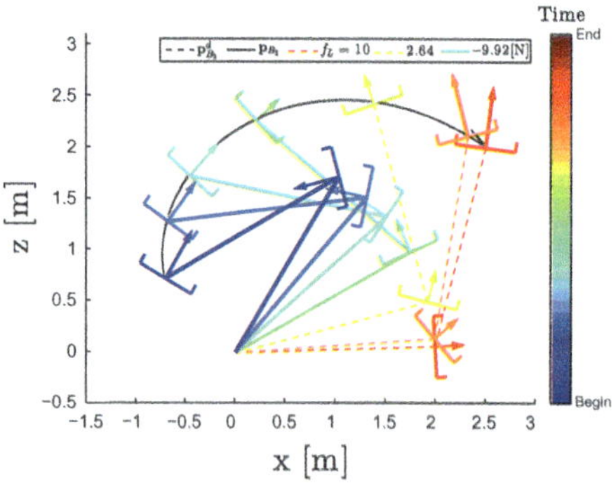

(c) Trajectory visualization from time Begin = 0 [s] to End = 6 [s]. The thrust arrow is drawn thicker and longer when the thrust intensity is higher.

Fig. 7.4 Simulation results: point to point motion.

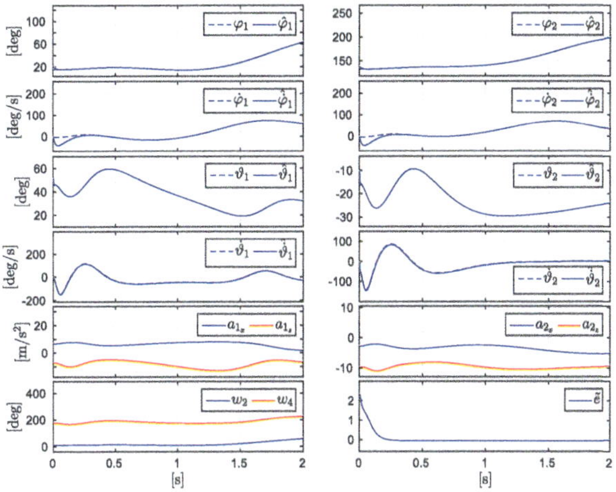

(a) Observer Results: the plot shows only the first 2 seconds of simulation.

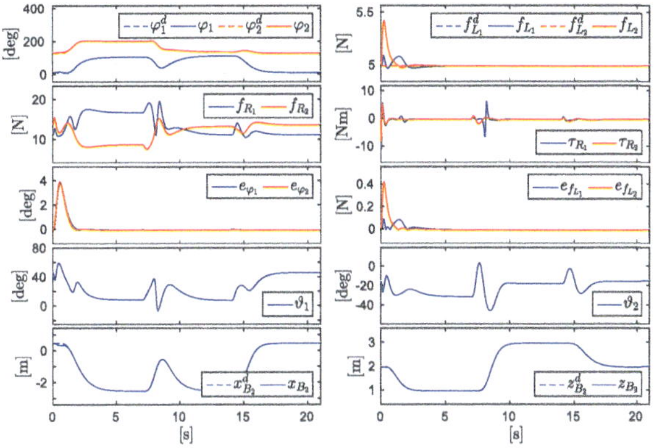

(b) Controller Results.

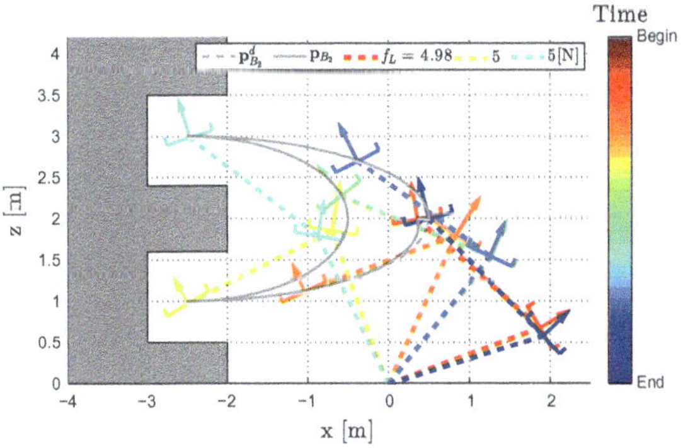

(c) Trajectory visualization from time Begin = 0 [s] to End = 21 [s].

Fig. 7.5 Simulation results: example of a search and rescue task.

shows a non zero tracking error, actually, as soon as the estimation error goes to zero, the outputs follow the desired trajectory with high fidelity during the remaining time of the simulation. An animation of the simulations is also available at [4].

References

1. Tognon, M., Franchi, A.: Nonlinear observer for the control of bi-tethered multi aerial robots. In: 2015 IEEE/RSJ International Conference on Intelligent Robots and Systems, pp. 1852–1857, Hamburg, Germany (2015)
2. Tognon, M., Franchi, A.: Control of motion and internal stresses for a chain of two underactuated aerial robots. In: 14th European Control Conference, pp. 1620–1625, Linz, Austria (2015)
3. Marino, R., Tomei, P.: Nonlinear Control Design: Geometric. Prentice Hall, Adaptive and Robust (1996)
4. video 4: Nonlinear observer for the control of bi-tethered multi aerial robots. https://www.youtube.com/watch?v=JdMAEneWolk (2015)

Chapter 8
Conclusions

8.1 Summary of the Book

With the aim of advancing in the control and motion planning for aerial robots interacting with the environment, in this book we extensively studied*tethered aerial vehicles*. These consists of unidirectional-thrust aerial vehicles connected to a moving or fixed point on the ground by a link whose length can be changed by a link actuator. For this general system, we produced a complete theoretical analysis of its dynamics and intrinsic properties, the controllability and the observability with a minimal set of standard sensors.

Starting with the investigation of the differential flatness of the systems, a very useful and powerful property of dynamical systems, we proved the existence of two flat outputs. The first, $\mathbf{y}^a$, is directly linked to physical interaction. It contains the position of the vehicle with respect to the anchoring point, the rotation along the thrust vector (standard flat output for an unidirectional-thrust vehicle in free-flight) and the *internal force* along the link. This tells us that the position of the aerial vehicle and the interaction force between the robot, the link and the system at the other end, can be controlled independently. Thanks to the generality of the computed dynamic model, the internal force can be tension and/or compression accordingly to the specific implementation. The second flat output, $\mathbf{y}^b$, contains the position of the vehicle with respect to the anchoring point, the rotation along the thrust vector and the angle ϑ_A that is related to the *attitude of the vehicle* with respect to the link. The latter output entry is unusual for unidirectional-thrust aerial vehicles because, in the free-flight configuration, the attitude is (in terms of differential flatness) a by-product of the translational motion. This adds new potential capabilities to the system.

Aiming to control those outputs and not only the position of the vehicle like in the majority of the state of the art, we designed and experimentally validated two first *hierarchical controllers*. Those are based on the separation principle between translational and rotational dynamics and exploit the flatness to compute the feed-

M. Tognon and A. Franchi, *Theory and Applications for Control of Aerial Robots in Physical Interaction Through Tethers*, Springer Tracts in Advanced Robotics 140,
https://doi.org/10.1007/978-3-030-48659-4_8

forward terms. The conducted experimental tests proven the validity of the method in quasi-static conditions but also shown its limitations when asked to track highly dynamic trajectories. For the goal of precisely tracking any sufficiently smooth time-varying trajectory (not only stabilization like in the state of art), we used the dynamic feedback linearization method to design a second pair of controllers. Parts of those control results were also extended to the particular case of a passive link actuator, which is an interesting case thanks to its simplicity. However, in this case, the internal force along the link is not controllable anymore, and neither the attitude-related variable.

All controllers require the full knowledge of the state of the system in order to compute the control action. Though, in practice it is difficult or even impossible to directly measure the full state of the system. Thus, motivated by the practical and theoretical relevance of the problem, we investigated which is the minimal set of standard sensors that make the system observable. Assuming that the motion of the anchoring platform is known, we proved that the standard onboard IMU together with three encoders measuring the attitude and the length of the link, are enough to obtain an estimation of the full dynamic state (including, e.g., the generalized velocities of the system). Differently from the state of the art in which an observer based on a quasi-static assumption was proposed, we aimed to design an almost globally convergent state estimator. For this purpose, we found some nonlinear transformations of the measurements that bring the system in the canonical controllability form. This allowed us applying a high gain observer. The case in which the link has a constant length and the vehicle is constrained on a vertical 2D plane is of particular interest. Under these assumptions, we proved that only onboard IMU alone is enough to retrieve an estimation of the full state (including position, attitude and generalized velocities). Such a surprising result is rather unique in the panorama of state estimation for aerial vehicles, indeed a positional measurement is typically always needed. Also in this case, we found some nonlinear transformations of the measurements and of the state such that to apply a high gain observer.

From the theoretical study, we passed to a more applicative and practical problem: *landing and takeoff of an unidirectional-thrust aerial vehicle on/from a sloped surface*. This problem is normally very challenging in a free-flight configuration due to the underactuation and the consequential need of a precise motion planning and tracking. On the other hand, we proved that the use of the tether makes those maneuvers much more safe, reliable and robust to model uncertainties and tracking errors. For the practical execution, we successfully employed the hierarchical controller for $\mathbf{y}^b$. A motion planner based on an optimal control method was also designed to improve the reliability and smoothness of the landing and takeoff.

Finally, we considered a multi-robot extension composed of two aerial vehicles tethered by two links to the ground and to each other, forming a chain-like system. The system is of particular interest for its similarity with a planar two-link manipulator, where the actuators are aerial vehicles. For this system, we extended the flatness and the dynamic feedback linearizability with respect to $\mathbf{y}_2^a$ (containing the position and internal link force for both couples vehicle/link). Also in this case, we designed a high gain observer based on IMU and encoders measurements.

8.2 Future Works

We presented a complete theoretical analysis of tethered aerial vehicles. Nevertheless, some additional works could be done, e.g., formally proving the stability of the proposed hierarchical controller, proven only experimentally. Other interesting extensions could be done toward an automatic identification of the system parameters and an adaptive controller, or toward a cooperative control with the moving platform. Another work could be to consider the problem of landing (and takeoff) on a surface with variable attitude, like a ship in a rough see. From the control and motion planning point of view, an interesting problem could be the one of using a tethered robot for the exploration of unknown and cluttered environments. On the other hand, given the level of maturity of the theoretical part, many are works that could be done form the practical and engineering point of view. Firstly, we plan to experimentally test the proposed dynamic feedback linearizing controller and the observers. Those could be then employed to enhance the performance during the landing and takeoff maneuvers in highly dynamic cases. However, for the real application, some improvements of the system have to be considered. For example a small winch could be added to unroll and roll-up the cable immediately before and after the tethered maneuvers. A more suitable anchoring mechanism can be designed according to the type of landing surface. Finally, the robot could be equipped with an onboard vision system to identify the position of the anchoring and landing points, as well as the landing surface attitude.

The manufacturer's authorised representative in the EU is Springer Nature Customer Service Centre GmbH, Europaplatz 3, 69115 Heidelberg, Germany. If you have any concerns regarding our products, please contact ProductSafety@springernature.com

Printed and bound by CPI Group (UK) Ltd, Croydon, CR0 4YY
30/09/2026
02225665-0012